SURFACE ANALYSIS AND PRETREATMENT OF PLASTICS AND METALS

Edited by

D. M. BREWIS, B.Sc., Ph. D.

School of Chemistry, Leicester Polytechnic, Leicester, UK

APPLIED SCIENCE PUBLISHERS
LONDON and NEW JERSEY

APPLIED SCIENCE PUBLISHERS LTD
Ripple Road, Barking, Essex, England
APPLIED SCIENCE PUBLISHERS, INC.
Englewood, New Jersey 07631, USA

British Library Cataloguing in Publication Data

Surface analysis and pretreatment
of plastics and metals
1. Materials   2. Surfaces (Technology)
I. Brewis, D. M.
620.1'1   TA403.6

ISBN 0-85334-992-4

WITH 28 TABLES AND 84 ILLUSTRATIONS

Photoset in Malta by Interprint Ltd
Printed in Great Britain by Galliard (Printers) Ltd, Great Yarmouth

SURFACE ANALYSIS AND PRETREATMENT OF PLASTICS AND METALS

PREFACE

To achieve satisfactory adhesion for printing, coating and bonding operations, it is necessary to pretreat many important plastic and metallic substrates. For example, the largest single application of plastics, i.e. packaging, often involves pretreatment to achieve satisfactory print and coating adhesion. Another example of increasing importance is the bonding of metals as an alternative to mechanical fastening techniques. Again the correct pretreatment is necessary, especially in aerospace and other demanding applications.

A satisfactory pretreatment must not only give the appropriate initial level of adhesion but must also provide 'durability', i.e. the adhesion must not be seriously affected by environmental conditions, in particular elevated temperatures, water and ions.

In order to understand the benefits provided by pretreatments, and ultimately to provide better pretreatments, it is necessary to study the physical and chemical changes caused by the pretreatments. This book provides descriptions with many examples of the techniques that may be used, including the latest spectroscopic methods. Some of these techniques may also be used to determine the locus of failure in a bonded structure or coated substrate; such information is important when seeking improved adhesion. A parallel approach in the study of pretreatments is provided by surface thermodynamics, and this subject is also examined in detail.

The use of modern analytical techniques and thermodynamics is highlighted in four important 'case studies', namely steel, aluminium, polyolefins and polytetrafluoroethylene. It is hoped that the detailed studies of these substrates will provide the reader with a good overall view of surface treatments, and that the approaches used and complications observed will be useful to those studying other substrates.

D. M. BREWIS

CONTENTS

LIST OF CONTRIBUTORS

W. L. BAUN, B.Sc., F.A.I.C.

Mechanics and Surface Interactions Branch, Materials Laboratory (AFWAL/MLBM), Wright Patterson Air Force Base, Ohio 45433, USA.

D. M. BREWIS, B.Sc., Ph.D.

School of Chemistry, Leicester Polytechnic, PO Box 143, Leicester LEI 9BH, UK

D. BRIGGS, B.Sc., Ph.D.

Petrochemicals and Plastics Division, Imperial Chemical Industries Ltd., PO Box 6, Bessemer Road, Welwyn Garden City, Hertfordshire AL7 1HD, UK.

A. B. CHRISTIE, B.Sc., A.R.C.S., Grad.R.S.C.

Department of Physics, Loughborough University of Technology, Loughborough, Leicestershire LE11 3TU, UK.

B. C. COPE, A.P.R.I., Ph.D.

School of Chemistry, Leicester Polytechnic, PO Box 143, Leicester LEI 9BH, UK.

R. H. DAHM, B.Sc., Ph.D.

School of Chemistry, Leicester Polytechnic, PO Box 143, Leicester . LEI 9BH, UK.

A. C. MOLONEY, B.Sc., Ph.D.

École Polytechnique Fédérale de Lausanne, Laboratoire de Polymères, 32, ch. de Bellerive, CH-1007 Lausanne, Switzerland.

D. G. RANCE, B.Sc., Ph.D.

Petrochemicals and Plastics Division, Imperial Chemical Industries Ltd, PO Box 6, Bessemer Road, Welwyn Garden City, Hertfordshire AL7 1HD, UK.

J. M. SYKES, M.A., Ph.D., C.Eng., C.Chem., M.R.C.S., M.I.M., M.I.Corr.T.

Department of Metallurgy and Science of Materials, University of Oxford, Parks Road, Oxford OX1 3PH, UK.

J. M. WALLS, B.Sc., Ph.D., M.Inst.P., M.I.M., C.Eng.

Department of Physics, Loughborough University of Technology, Loughborough, Leicestershire LE11 3TU, UK.

Chapter 1

INTRODUCTION

D.M. Brewis
Leicester Polytechnic, Leicester UK

1. OUTLINE

The success of many products is dependent on adequate adhesion between two or more materials. Some examples are given in Table 1. In order to achieve adequate adhesion for the lifetime of the product it is often necessary to carry out a surface pretreatment which may be mechanical or chemical. The success of a pretreatment must be judged not only by the initial adhesion but also by its durability, a topic discussed in Chapters 7 and 8.

The surface treatments for some important substrates are discussed in detail in later chapters. However, before the success of these pretreatments can be understood, the mechanisms of adhesion must be considered, together with the factors that affect adhesion.

2. THEORIES OF ADHESION

There is much controversy regarding the basic reason for the adhesion that exists between different materials. The four main mechanisms are now briefly reviewed; they are considered mainly from the viewpoint of adhesives, but many of the considerations also apply to coatings.

2.1. Mechanical theory
According to this theory, the adhesive interlocks around the irregularities or pores of the substrate. However, when the adhesion to a rough surface is concerned, other factors must be considered. A rough surface will have

TABLE 1

EXAMPLES WHERE GOOD ADHESION IS REQUIRED

Substrate–adhesive–substrate
Substrate–printing ink
Substrate–extrusion coating, e.g. polyethylene on aluminium
Substrate–paint
Substrate–metal, e.g. metallised polypropylene
Composites, e.g. glass fibre reinforced plastics
Plastic–plastic (heat sealing)

a larger potential bonding area than a smooth one. On the other hand, stress concentrations due to voids may be important.[1] Kinloch[2] has emphasised that the energy dissipated viscoelastically and plastically during fracture may be much larger with a rough surface. Therefore, if roughness is increased by a surface treatment, the reasons for the changes in adhesion may be complex.

Borroff and Wake[3] showed that the adhesion between various adhesives and cotton was largely unaffected by surface treatments of the cotton. This is in agreement with the idea of mechanical keying.

The metal plating of the polymer acrylonitrile–butadiene–styrene (ABS) is probably an example of mechanical adhesion. To obtain satisfactory adhesion it is necessary first to carry out a pretreatment with chromic acid, which dissolves rubber particles near the polymer surface leaving a porous structure. A metal may then be deposited from solution into the porous structure providing a mechanical key. Atkinson et al.[4] have produced some electron micrographs which support this mechanism although the factors noted above must be considered.

Packham and co-workers have provided much evidence showing the importance of surface roughness in the adhesion between polyethylene and metals; for example, see references 5 and 6.

2.2. Adsorption theory

The adhesive macromolecules are adsorbed on to the surface of the substrate and are held there by various forces of attraction. The adsorption is usually physical, i.e. due to van der Waals' forces. However, hydrogen bonding and primary bonding (ionic or covalent) are involved in some cases. If primary bonds are involved the term chemisorption is used. This theory assumes a definite interface between the adhesive and adherend.

Tabor[7] and others have shown that dispersion forces, which are one of three van der Waals' interactions, alone are more than sufficient to account for the highest joint strengths observed. The discrepancy between the theoretical and practical values is probably due to defects such as voids. In any case chemical bonding will lead to much better joint strength durability, which is a major problem with adhesive bonded metal joints.

Hydrogen bonding is believed to be important in the bonding of tyre cords (nylon[8] or polyester[9]) to rubber. There is good evidence[10,11] that hydrogen bonding is involved in the self-adhesion of corona-treated polyethylene (see Chapter 9).

The direct evidence of chemical bonding in adhesion is limited in the main to silane primers and coupling agents. Gettings and Kinloch[12] have provided strong evidence for chemical bonding between a silane primer and steel. Using secondary ion mass spectroscopy they detected the presence of $FeOSi^+$ ions; however, Baun points out in Chapter 3 (section 3) that the bombarding ions can cause the formation of molecular complexes.

Although the exact nature of the interactions at the interface may be uncertain, the adsorption theory of adhesion is the most widely accepted mechanism except in the USSR.

2.3 Diffusion theory

The adhesive macromolecules diffuse into the substrate, thereby eliminating the interface. The theory can therefore only apply to polymeric substrates. It requires that the macromolecules of the adhesive and adherend have sufficient chain mobility and that they are mutually soluble.

Voyutskii[13] and Vasenin[14] in particular have been strong advocates of the diffusion mechanism. Voyutskii's experimental evidence is based mainly on autohesion experiments, i.e. bonding experiments where the adhesive and substrate are identical. In particular, he studied the bonding of rubbers at elevated temperatures and found that the joint strength increased with

(a) increased period of contact,
(b) increasing temperature,
(c) increasing pressure,
(d) decreasing molecular weight,
(e) addition of plasticisers,

and decreased with cross-linking.

However, although diffusion undoubtedly occurs when two identical (or similar) polymers are brought together at relatively high temperatures (i.e. autohesion), the evidence is equally consistent with the adsorption theory, since the factors (a)–(d) all affect the degree of contact achieved between the adhesive and substrate.

There is little direct evidence of diffusion in adhesive joints although various workers, including Bueche *et al.*,[15] have demonstrated self-diffusion in the bulk phase. However, there is no doubt that diffusion takes place during the solvent welding of plastics, e.g. when two sheets of poly(vinyl chloride) are softened with a chlorinated solvent and then joined. Diffusion will also take place when two pieces of the same plastic are heat sealed.

2.4. Electrostatic theory

The chief proponents of the electrostatic or electronic theory have been Derjaguin and his co-workers.[16] They proposed that adhesion was due to electrostatic forces, arising from the transfer of electrons from one material of an adhesive joint to another. Evidence put forward in support of this theory includes the observation that the parts of a broken adhesive joint are sometimes charged. However, this does not prove that the strength of the joint was due to electrostatic charges because the latter may have been produced during the destruction of the joint.

Derjaguin and Smilga[16] also claim that peeling forces are often very much greater than can be accounted for by van der Waals' forces or chemical bonds. They also claim that the strong dependence of peel strength on testing speed cannot be explained in terms of these forces but can by means of the electrostatic theory. However, Schonhorn[17] and others point out that most of the work done in a peeling experiment is due to deformation of the materials comprising the joint rather than to overcoming the molecular forces across the interface. The rate dependence of peel strength can also be explained in terms of rate dependent mechanical properties of the adhesive and substrate.

There is some independent evidence for the electrostatic theory. For example, Weaver[18] found that the adhesion between copper and poly(methyl methacrylate) fell after exposure to a glow discharge in a vacuum chamber for a few minutes.

2.5. Conclusions

There are probably cases where each of the four mechanisms is dominant. As far as the systems in the present book are concerned, the adsorption mechanism will be generally applicable but mechanical keying and diffusion will sometimes play an important role.

3. FACTORS AFFECTING ADHESION

In those cases relevant to this book, adhesion is usually achieved by the flow of a polymer into the irregularities of a solid surface. This is true with adhesives or coatings where the polymer may be in the form of a solution, dispersion or melt.

The basic requirements for good adhesion are:

(a) good contact between polymer and substrate,
(b) no weak layers at the interface,
(c) the polymer should possess appropriate mechanical properties.

Surface treatments can clearly affect the first two requirements. The mechanical properties of the adhesive (or coating) are not directly relevant to this book.

Unfortunately it is not known what degree of contact is required for good adhesion. The degree of contact achieved depends on the viscosity of the adhesive or coating, and also on the surface energies of the wetting liquid and the substrate. The role of surface energies, which is discussed in detail in Chapter 6, is vital to our understanding of adhesion. However, perfect wetting would be pointless if a region of low strength existed in the structure. There are many potential weak boundary layers with both metallic and plastic substrates,[19] although there is much controversy with respect to the frequency of occurrence of these weak layers.[20] It is likely that regions of low strength on substrates are often displaced by adhesives or coatings and they therefore do not represent weak boundary layers.[20]

Many of the factors that affect the strength of an adhesive joint are given in Table 2; most of these factors are also relevant to the adhesion between a coating and the substrate. Obviously only the nature of the substrate is directly affected by a pretreatment. However, many of the other factors are related to the nature of the surface, e.g. if the chemistry of the surface is changed, the interaction with the adhesive will vary.

TABLE 2
FACTORS AFFECTING ADHESION

1. *Nature of substrate surface*
    (a) Surface geometry
    (b) Surface chemistry
    (c) Mechanical strength of surface regions
2. *Mechanical properties of substrate*
3. *Nature of adhesive*
    (a) Viscosity
    (b) Chemical nature
    (c) Mechanical properties
4. *Bonding conditions*
    (a) Pressure
    (b) Temperature
    (c) Time
5. *Conditions prior to testing*
    (a) Temperature
    (b) Humidity
    (c) Stress
6. *Testing conditions*
    (a) Strain rate
    (b) Temperature

4. EFFECTS OF SURFACE TREATMENTS

The effects that may result from a surface treatment are summarised in Table 3. They are discussed in detail in later chapters but at this stage a few general points are made.

Mechanical pretreatments, e.g. sand blasting, often cause much roughening of a surface. The effect on adhesion is complex. There is an increase in potential bonding area and also mechanical keying can occur. However, if the viscosity of the adhesive or coating is high, or the time the wetting liquid is in its fluid state is short, then the degree of contact could actually be less after the roughening; this would be especially true with deep narrow pits. The resulting voids would also cause stress concentrations. However, provided the viscosity of the adhesive is fairly low, roughening the surface will probably lead to an increase in adhesion.

The solvent etching of a polymer may have a similar effect to a mechanical treatment of a metal. Garnish and Haskins[21] showed that trichloroethylene vapour caused very rapid and extensive roughening of

TABLE 3
EFFECTS OF SURFACE PRETREATMENTS

Treatment	Possible effects on surface
Mechanical	(a) Remove weak boundary layer (P, M)[a]
	(b) Increase surface roughness (P, M)
Solvent etch	(a) Remove weak boundary layer (P, M)
	(b) Weaken surface region by plasticisation (P)
	(c) Increase surface roughness (P)
Chemical	(a) Remove weak boundary layer (P, M)
	(b) Increase (or decrease) surface roughness (P, M)
	(c) Alter surface chemistry with consequent changes in the rate and degree of wetting (P, M)

[a] Abbreviations: P, plastic; M, metal.

a polypropylene surface. They found that the adhesion of the polypropylene to an epoxide adhesive increased about six-fold. However, the possibility that the surface regions of a polymer will be weakened by plasticisation should not be ignored.

Chemical pretreatments can cause even more complex changes than mechanical pretreatments. Chemical treatments often roughen the surface, e.g. anodising results in a very porous surface. Occasionally a surface treatment may smooth a surface: this is thought to be the case with certain levels of corona treatment of polyolefins. The changes in surface geometry and chemistry will affect the rate and degree of wetting. The change in surface chemistry will also alter the degree of interaction between the adhesive (or coating) and the substrate. It is well established that chemical and mechanical pretreatments can remove weak layers on metal surfaces, e.g. weak oxides, but whether pretreatments act in this way with polymers is the subject of much controversy.[20]

In order to understand the effects of surface treatments it is necessary to study in detail these physical and chemical changes.

5. STUDY OF CHANGES CAUSED BY PRETREATMENTS

Topographical and morphological information is usually provided by optical and electron microscopy (Chapter 5). These techniques are therefore very useful to follow the physical changes due to the pretreatments. Microprobe analysis is often used in conjunction with electron microscopy to study any changes in elemental distribution.

In recent years a wide range of analytical techniques has become available to study the surface chemistry of metals. The most useful techniques are: Auger Electron Spectroscopy, AES; Electron Spectroscopy for Chemical Analysis, ESCA (alternatively called X–ray Photoelectron Spectroscopy, XPS); Ion Scattering Spectroscopy, ISS; and Secondary Ion Mass Spectroscopy, SIMS. The techniques of electron spectroscopy are discussed in Chapter 2 and those of ion spectroscopy in Chapter 3. It is possible to obtain quantitative information on chemical composition and its variation with depth.

ESCA (or XPS) is also very useful in the study of polymer surfaces, but AES causes excessive damage to such surfaces. Infrared techniques (Multiple Internal Reflection IR and Fourier Transform IR) are also useful in the study of pretreatments for plastics. The chemical analysis of polymer surfaces is discussed in Chapter 4.

The use of contact angle measurements to characterise surfaces is discussed in Chapter 6. The use and limitations of the various thermodynamic treatments are examined.

It is important to identify the locus of failure in an adhesive joint. If interfacial failure occurs, a better surface pretreatment should lead to a better joint performance. Often failure occurs very near the surface, i.e. a very thin layer of adhesive remains on the surface. To the naked eye the failure appears interfacial (also termed adhesive failure) but several of the above techniques could be used to show that cohesive failure had occurred (see Chapter 3).

6. PRETREATMENTS FOR METALS

The surfaces of the commonly used metals are often contaminated with rolling lubricants or incidental contaminants. In addition the metal may be covered by a weak oxide layer. For some applications it is sufficient just to degrease and abrade, but to obtain maximum strength, reproducibility and durability, a chemical pretreatment will be necessary. The treatment must remove any weak oxide layer and replace it with a strong oxide of controlled thickness, morphology and topography.

Much of the work on pretreatment of metals prior to adhesive bonding has been for the aircraft industry, where adhesives are being increasingly used in primary and secondary structures. Therefore most work has been carried out on aluminium and its alloys, and to a lesser extent on magnesium and titanium. The usual methods for pretreating aluminium are a chromic acid pickle,[22] chromic acid anodising[23] or phosphoric acid anodising.[24] Not only is the initial joint strength of great importance, but also their durability. Cotter[25] has demonstrated large differences in the resistance of various surface treatments to high humidity. Pretreatments for aluminium are discussed in detail in Chapter 8.

The adhesion problems between steel and organic polymers are often centred on coatings rather than adhesives, an extremely important example being the painting of road vehicles and ships. Surface treatments for steel are discussed in detail in Chapter 7. Procedures for preparing other metal surfaces for bonding and coating are given elsewhere.[26]

7. PRETREATMENTS FOR PLASTICS

The growth in the consumption of plastics has been extremely high over the last 30 years. Packaging is the most important single application of plastics, using about 25% of total production. Adhesion problems in bonding, coating, heat sealing and printing operations have been numerous.

To achieve satisfactory adhesion to a polymeric substrate, a pretreatment is often necessary. With some polymers a solvent wipe or mild abrasion may be sufficient to give the necessary adhesion. However, with other polymers it may be necessary to modify the surface chemically.

Polyolefins are the most widely used synthetic polymers[27] and to achieve satisfactory adhesion to them it is usually necessary to carry out a pretreatment. It was for these reasons that polyolefins were selected for detailed consideration in Chapter 9. Whereas surface treatments for metals are normally carried out in acid solution, the two most widely used treatments for polyolefins are dry processes, namely the corona discharge treatments for films and the flame treatment for thicker sections.

Although polytetrafluoroethylene is a low tonnage plastic, it is of considerable technological importance and it is well-known for its 'non-stick' properties. Further, its adhesion characteristics are especially complex and for these reasons polytetrafluoroethylene has been selected for detailed consideration in Chapter 10.

8. ASSESSMENT OF ADHESION

The term adhesion has two distinct meanings. To the physical chemist it is associated with intermolecular forces acting across an interface, and involves a consideration of surface energies and interfacial tensions (Chapter 6). To those involved with the technology of bonding, coating and printing the term adhesion has another meaning, e.g. in adhesive bonding it is the mechanical force required to separate the substrates that are bonded with the adhesive. In terms of this second meaning there are various ways of assessing the level of adhesion depending on the technology involved. With adhesives, the butt, lap or peel tests are the most commonly used, but there are many more.[28] Again, with coatings and printing inks, there are special tests to assess the adhesion.[29]

9. CONCLUSIONS

Pretreatments of plastics and metals are often necessary to give satisfactory long-term results for bonding, coating and printing operations. In order to understand the reasons for the beneficial effects, it is necessary to study the physical and chemical changes brought about by the pretreatments. The most important techniques available to carry out such a study are described in Chapters 2–6.

Different substrates present their own complications as illustrated in Chapters 7–10. However, a consideration of these well-studied materials should be useful in the understanding of other substrates.

REFERENCES

1. Wake, W. C. in *Adhesion*, D. D. Eley (Ed.), Oxford University Press, London 1961, p. 191.
2. Kinloch, A. J., *J. Mat. Sci.*, **15** (1980), 2141.
3. Borroff, E. M. and Wake, W. C., *Trans. Inst. Rubber Ind.*, **25** (1949), 190.
4. Atkinson, E. B., Brooks, P. R., Lewis, T. D., Smith R. R. and White K. A., *Plast. Inst. Trans. J.*, **35** (1967), 549.
5. Malpass, B. W., Packham, D. E. and Bright, K., *J. Appl. Polym. Sci.*, **18** (1974), 3249.
6. Evans, J. R. G. and Packham, D. E., *J. Adhesion*, **10** (1979), 177.
7. Tabor, D., *Rep. Progr. Appl. Chem.*, **36** (1951), 621.
8. Pritchard, W. H. in *Aspects of adhesion—6*, D. J. Alner (Ed.), University of London Press, London 1971, p. 11.

9. Mather, J., *Brit. Polym. J.*, **3** (March 1971), 58.
10. Owens, D. K., *J. Appl. Polym. Sci.*, **19** (1975), 265.
11. Briggs, D. and C. R. Kendall, *Polymer*, **20** (1979), 1053.
12. Gettings, M. and Kinloch, A. J., *Surf. Interface Anal.*, **1** (1980), 189.
13. Voyutskii, S. S. *Autohesion and adhesion of high polymers*, Interscience, New York 1963. Translation from the Russian by S. Kaganoff.
14. Vasenin, R. M. in *Adhesion fundamentals and practice*, UK Min. of Tech. (Ed.), Elsevier, London 1969, p. 29.
15. Bueche, F. J., Cashin, W. M. and Debye, P., *J. Chem. Phys.*, **20** (1952), 1956.
16. Derjaguin, B. V. and Smilga, V. P., *Proc. 3rd Internat. Congress of Surface Activity*, **11** (1960), 349.
17. Schonhorn, H. in *Adhesion fundamentals and practice*, UK Min. of Tech. (Ed.), Elsevier, London 1969, p. 12.
18. Weaver C., in *Adhesion fundamentals and practice*, UK Min. of Tech. (Ed.), Elsevier, London 1969, p. 46.
19. Brewis, D. M. in *Polymer science*, A. D. Jenkins (Ed.), North-Holland Publishing Co., London 1972, p. 934.
20. Brewis, D. M. and Briggs, D., *Polymer*, **22** (1981), 7.
21. Garnish, E. W. and Haskins, C. G. in *Aspects of adhesion—5*, D. J. Alner (Ed.), University of London Press, London 1969, p. 259.
22. UK Ministry of Defence, Defence Standards 03-2/1.
23. UK Ministry of Defence, Defence Standards 151.
24. McMillan, J. C., Quinlivan, J. T. and Davis, R. A., *SAMPE Quarterly*, **6** (1976), 3.
25. Cotter, J. L. in *Developments in adhesion—1*, W. C. Wake (Ed.), Applied Science Publishers Ltd, London 1977, p. 1.
26. Code of Practice for Cleaning and Preparation of Metal Surfaces, CP 3012: 1972, British Standards Institution.
27. *Eur. Plast. News*, (January 1981). 6.
28. Shields, J. T. *Adhesives handbook*, 2nd edn., Newnes–Butterworth, London 1976.
29. Mittal, K. L., *Adhesion measurement of thin films, thick films and bulk coatings*, ASTM STP 640.

Chapter 2

SURFACE ANALYSIS BY AUGER AND
X-RAY PHOTOELECTRON SPECTROSCOPY

J. M. WALLS and A. B. CHRISTIE
Loughborough University of Technology, Loughborough, UK

1. INTRODUCTION

A number of techniques are now available for measuring the composition of any solid surface. Since the surface plays such a crucial role in many processes such as oxidation, discolouration, wear and adhesion, these techniques have established their importance in a number of scientific, industrial and commercial fields. The most widely used techniques for surface analysis are Auger Electron Spectroscopy (AES), X-Ray Photoelectron Spectroscopy (XPS) or Electron Spectroscopy for chemical Analysis (ESCA), and Secondary Ion Mass Spectrometry (SIMS). Each of these techniques has the ability to determine the composition of the outermost atomic layers, although each technique possesses its own special advantages and disadvantages. This chapter introduces the first two of these techniques, AES and XPS, and the ways in which they can be usefully applied. Since both of these techniques involve electron spectroscopy, the concepts, methods and instrumentation are similar. In contrast, SIMS is essentially a technique based on mass spectrometry and for this reason it is dealt with separately in Chapter 3.

AES and XPS are generally described as 'surface analysis' techniques, but this term can be misleading. Although they derive their usefulness from their intrinsic surface sensitivity, they can also be used to determine the composition of much deeper layers. This is normally achieved by the controlled erosion of the surface by ion bombardment or by tapering the surface by some mechanical means. In this way, composition–depth profiles can be obtained which provide a powerful means of analysis for thin films, surface coatings and their interfaces. Clearly, this capability also makes AES and XPS ideal for the study of adhesion and adhesives.

Both AES and XPS have already been applied in some areas to the study of adhesives and surface pretreatments although the full potential of the techniques has yet to be realised. Both techniques are ideal for the characterisation of surface contamination. The need for such analysis is clear since the development of surface pretreatments, in itself, confirms the importance of surface contamination in subsequent bond performance. Indeed the surface composition of adherends following preatreat-ment also requires surface analysis, since processes such as acid etching, solvent degreasing or grit blasting do not leave atomically clean surfaces, but often merely substitute more acceptable forms of contamination. The provision of a composition–depth profile through the bond itself can be used to characterise the bond chemically, and the results can be cor-related to bond performance. The environment of the bond, and its handling, may also affect durability and strength. All these problems, together with analyses to determine the precise locus of failure, can now be approached in a scientific way using surface analytical methods.

2. BASIC PRINCIPLES

2.1. Auger electron spectroscopy (AES)

Auger electrons are produced by bombarding the specimen surface with low energy electrons (1–10 keV). Some of the atoms within the sample are ionised and Fig. 1 illustrates the kind of electron re-arrangement which can take place within the atom as a result. Thus, if, for example, an electron from the core level is ejected, an electron from the L_2 level may fill the vacant site releasing an amount of energy $E_K - E_{L_2}$. This energy may now be transferred to another electron possibly in the L_3 level, which is then ejected from the solid. This latter electron is the Auger electron and its kinetic energy E is approximately given by

$$E \simeq E_K - E_{L_2} - E_{L_3} - \phi$$

where ϕ is the work function of the energy analyser.

Since E_K, E_{L_2} and E_{L_3} are all characteristic of the particular element concerned, it is possible, by measuring the energies and the number of Auger electrons, to determine the chemical composition of the sample surface. The technique is sufficiently sensitive to enable detection of as

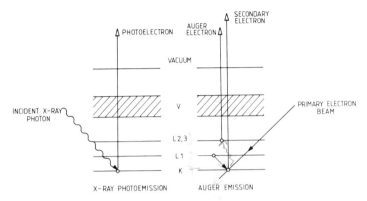

FIG. 1. A schematic energy level diagram which illustrates the photoexcitation of a K shell electron by a low energy X-ray photon ($h\nu$) and the transition leading to the emission of a KL_1, $L_{2,3}$ Auger electron.

little as 0·1% of a monolayer of impurity in the surface. Only hydrogen and helium cannot be detected since these elements possess insufficient energy levels for the Auger transition to occur.

When the primary electron beam interacts with the sample surface, a number of mechanisms exist whereby electrons are reflected, scattered or ejected from the surface to give the secondary electron distribution $N(E)$ with energy (E) at all values from zero up to the primary beam energy. The Auger features in this energy distribution are comparatively small in amplitude and occur on a large sloping background which makes their interpretation difficult. Consequently, the spectra are usually differentiated electronically so that $dN(E)/dE$ versus E is displayed.[1] This suppresses the background and accentuates the prominence of the Auger peaks.

An Auger spectrum in the $dN(E)/dE$ form for a rubber-to-brass interface is shown in Fig. 2(a). By convention, the energy of the Auger peak is taken to be the energy corresponding to the minimum of the negative part of the Auger transition since this is usually the most sharp and prominent. Different elements are identified by the presence of characteristic peaks at well-defined energies. In practice, the peaks can be assigned by determining their energy and using a chart of Auger transitions or, alternatively, the various peaks can be identified by comparison with a compilation of standard spectra.[2,3] Peak overlaps can occur in spectra obtained from complex surfaces with large numbers of elements present, especially those from adjacent positions in the periodic

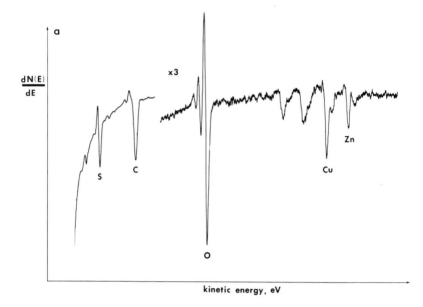

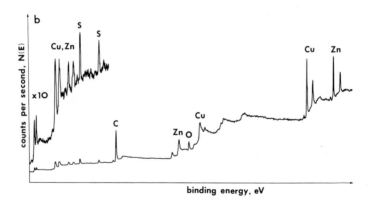

FIG. 2. (a) AES spectra from a rubber-to-brass interface, illustrating the high sensitivity to low atomic number elements. Primary beam energy 3keV, primary beam current $2\,\mu A$. (b) XPS spectrum from a rubber-to-brass interface. The sulphur photoelectron peak position is indicative of inorganic sulphide. X-ray energy $1486.6\,eV$ (AlK_α), analyser resolution $0.5\,eV$.

table. However, since most elements have more than one Auger transition, it is usually possible to confirm their presence using subsidiary peaks except when their concentration approaches the sensitivity limit. A particular problem can occur with those elements having only one Auger transition in the energy range used, but, even in the most difficult case of peak superposition, their presence can usually be inferred from the precise shape of the composite peak.

2.2. X-ray photoelectron spectroscopy (XPS)

In XPS, the sample surface is illuminated with a source of soft X-rays, photoionisation of a core level takes place and the resultant photoelectrons have a kinetic energy E which is related to the X-ray energy (hv) by the Einstein relation

$$E = hv - E_B - \phi$$

where E_B is the binding energy of the electron in the material and is characteristic of the individual atom. For any given electron shell, the K shell for instance, the electron binding energy increases with increasing atomic number as shown in Fig. 3. Thus, information of the binding energies of electrons within a sample allows direct qualitative analysis. Since the energy levels occupied by electrons are quantised, the photoelectrons have a kinetic energy distribution $N(E)$, consisting of a series of discrete bands that reflects the shell form of the electronic structure of the sample. In addition, the electron binding energies within any one element are not fixed and small variations of up to 10 eV may occur. These 'chemical shifts' are caused by changes in the valence electronic structure (chemical environment) of the atoms concerned and so may be used to provide information on chemical bonding.

An XPS spectrum taken from a rubber-to-brass interface is shown in Fig. 2(b).

2.3. Surface sensitivity

Although the incident radiation in both AES and XPS penetrates deep into the sample, it is only those electrons which are generated close to the outermost surface which escape with all their original energy intact. Those Auger or photoelectrons which originate further within the sample, lose energy in inelastic collisions and may no longer be identified as such. It is estimated that the mean free paths or electron escape depths (λ) range from 1 nm in metals to about 10 nm in organic coatings[4,5] and it is this which gives the techniques their great surface sensitivity. A

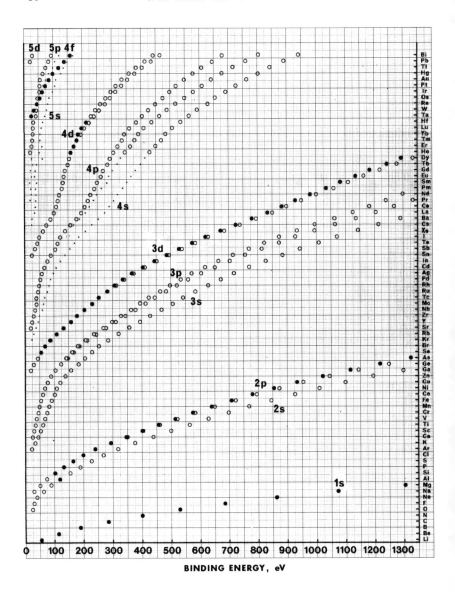

FIG. 3. Elemental core-level binding energies (in electron volts, eV) versus atomic number, for all core levels in the energy range 20–1350 eV. ●, Principal peak in wide (survey) scan; ○, other main photoelectron peaks; ·, peaks contributing < 2% to the total photoelectron intensity. From ref. 16.

compilation of escape depths is shown in Fig. 4 as a function of electron energy. This allows fine scale depth information to be obtained, since a comparison of the relative intensity of low- and higher-energy transitions can reveal changes in concentration of the elements within a few monolayers of the surface.

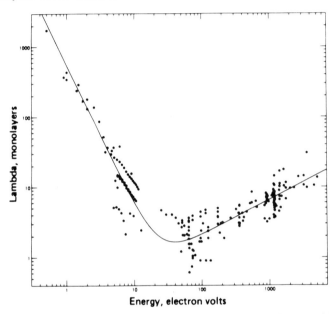

FIG. 4. A compilation of experimental measurements of inelastic mean free paths in monolayers, for elements (from Seah and Dench, ref. 5).

3. INSTRUMENTATION

An electron spectroscopy system may be divided conveniently into three parts, viz. an excitation source, an electron energy analyser, and an ultra-high vacuum system. This is shown schematically in Fig. 5. In addition to these basic requirements, most electron spectrometers also possess ancillary experimental facilities, which enhance their analytical capabilities, as well as an array of sophisticated electronics to aid data acquisition, processing and display.

3.1. The excitation source
The excitation sources available for electron spectroscopy may be di-

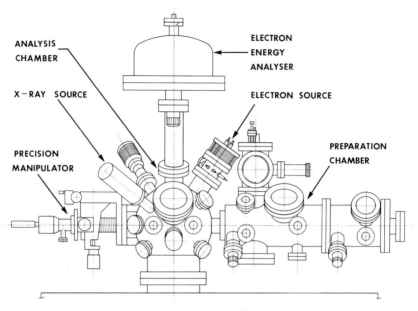

ANALYSIS
CHAMBER

ELECTRON
ENERGY
ANALYSER

X - RAY SOURCE

ELECTRON SOURCE

PRECISION
MANIPULATOR

PREPARATION
CHAMBER

FIG. 5. Schematic diagram of a combined ESCA/Auger electron spectrometer (courtesy V.G. Scientific Limited).

vided into two types: excitation sources for AES (electron sources) and excitation sources for photoelectron spectroscopy (ultraviolet and X-ray sources).

Electron sources in common use operate on the principle of thermionic emission. Electrons from a hot cathode (usually a tungsten filament, resistively heated to about 2500 K) are accelerated to between 1 and 10 keV before impact with the sample. The addition of focusing and deflection electrodes to this basic electron gun assembly can produce an electron beam of 0·5 μm diameter, which, when scanned or rastered across the sample surface, supplies information concerning elemental distribution and topography.

Ultraviolet sources can provide information about weakly bound electrons only (~ 20 eV), but their use is specialised and they are beyond the scope of this chapter. Ultraviolet photoelectron spectroscopy has been dealt with in detail elsewhere.[6]

Conventional X-ray sources operate on the principle of X-ray emission. Electrons from a tungsten filament are made to impinge upon a water-cooled solid anode, de-excitation of which produces X-rays of characteristic energies. The most commonly used targets are magnesium (characteristic K_α X-ray energy of 1253·6 eV) and aluminum (characteristic K_α X-ray energy of 1486·6 eV), which produce X-ray linewidths of 0·7 eV and 0·85 eV respectively. For some applications, it is desirable to use anode materials which produce characteristic X-rays of very high (e.g. CuK_α at 8048 eV) or very low (e.g. $ZrM\zeta$ at 151·4 eV) energy,[7] but such cases are relatively rare.

X-ray linewidths may be reduced with the aid of a monochromator, but the simultaneous loss of intensity which results means that mono-chromatised X-ray sources are used only in a very few applications. Use has also been made of synchrotron radiation as a tunable X-ray source, but its application is limited to certain aspects of fundamental solid state research.[8]

3.2. The sample

In principle, any type of sample, including gases[9] and liquids,[10] may be studied by electron spectroscopy, but limitations of instrument design usually restrict the range to low vapour pressure solid samples. Restrictions may also be imposed on sample size, but commercial electron spectrometers are presently available that will accept samples with dimensions of several centimetres. Early electron spectrometers were designed to hold only one, often specially machined, sample at a time,[11] but modern instruments are able to handle a number of powder, fibre or bulk solid samples simultaneously. Methods of sample mounting are numerous, and are often dictated by the nature of the sample itself. Bulk solid samples may be clipped, or fixed with double-sided adhesive tape. Powders can be brushed on to double-sided adhesive tape, but better results are obtained if the powder is simply contained in a small crucible. Fibres may be similarly mounted, or alternatively they may be clipped in position. In most spectrometers, the sample to be analysed is situated on a precision manipulator, such that the sample surface may be moved, relative to the excitation source, with a translational and rotational precision of 10 μm and 1° respectively.

3.3. The electron energy analyser

Following excitation by electrons or X-rays, the emitted Auger or photoelectrons are made to enter an electron energy analyser, which operates by dispersing the electrons in an electrostatic field according to their energy. The two types of electron energy analyser most commonly employed in electron spectrometers are the cylindrical mirror analyser

(CMA)[12] and the hemispherical analyser (HSA).[11] Although both types may be operated in either low or high resolution (high or low sensitivity) modes, the former design has the advantage of a relatively higher transmission at low resolution, whereas the latter has the advantage of a higher transmission at high resolution.[13]

Although high resolution (double-pass) CMAs, and high transmission HSAs, have been developed, the high sensitivity requirements of AES generally make the use of a CMA the most appropriate, while the requirements of high energy resolution in photoelectron spectroscopy make the HSA the best choice.

After leaving the analyser the energy-dispersed electrons are made to enter an electron multiplier, which may be operated in either a pulse counting (digital) or total current mode. By suitably scanning the analyser voltages, the output from the multiplier may be used to produce the electron spectrum.

3.4. The vacuum system

Most of the constituent parts of the electron spectrometer (electron gun, X-ray anode and electron multiplier) will only operate effectively in a high vacuum. In addition, the extreme surface sensitivity of electron spectroscopy necessitates the use of an ultra-high vacuum (UHV) system to maintain a clean sample surface. (Contaminants can build up at the rate of one monolayer per second in vacua as low as 10^{-6} Torr). Consequently, modern UHV systems are usually capable of producing base pressures of about 10^{-11} Torr, although pressures of about 10^{-9} Torr are generally acceptable for routine analysis. These low pressures are achieved using bakeable stainless steel systems and by the use of either oil diffusion pumps (usually polyphenyl ether), ion pumps, or turbomolecular pumping systems. The vacuum may be further improved with ancillary pumping systems such as titanium sublimation pumps and cryogenic traps. Many electron spectrometers are based on a single-chamber design, in which all of the sample mounting, experimentation and analysis takes place. Multiple-chamber spectrometers, with separate pumping lines to each chamber, have the advantage, however, of rapid sample introduction (via an entry lock) and independent sample preparation and analysis chambers (Fig. 6).

3.5. Experimental facilities

In many cases, a full characterisation of a sample surface can only be carried out with the aid of additional experimental facilities fitted to the

FIG. 6. General view of the multichamber arrangement on a combined ESCA/Auger electron spectrometer (courtesy V.G. Scientific Limited).

electron spectrometer. Such facilities may include ion guns, for *in situ* sample cleaning and the production of composition–depth profiles (described in detail in section 6); a sample heating and cooling stage; a sample fracture stage; an evaporation source; and gas handling facilities.

3.6. Data acquisition, processing and display

The signal emerging from the electron multiplier is amplified, and may then be processed either as a series of pulses (pulse counting mode—each pulse representing a single detected electron) or as an electric current (total current mode—proportional to the electron count rate). Pulse counting is only practical at low electron currents, and is therefore not generally used for AES, when the total current mode is necessary. In XPS and low current (high spatial resolution) AES, however, electron detection rates rarely exceed $3 \times 10^5 \text{s}^{-1}$, and pulse counting is possible.

Pulse-counted spectra are usually displayed directly as count rate versus electron energy (termed the $N(E)$ spectrum), whereas total current Auger spectra are usually displayed in differential form (the $dN(E)/dE$ spectrum) (see section 2). With the aid of numerical differentiation using a computer, pulse-counted Auger spectra may also be displayed in this form.

4. QUANTITATIVE INTERPRETATION

4.1. Basic principles

The rate of Auger or photoelectron emission (number of emitted electrons per second), from any given element, is proportional to the number of atoms of that particular element within the sampling volume. It is easily seen, therefore, that electron spectra may be used to obtain a quantitative elemental analysis of the sample under study, providing the relevant elemental sensitivity factors are known, or can be measured. Comprehensive reviews of the quantification of electron spectra have already been made by Powell and Larson,[14] and Seah,[15] and only a brief summary will be given here.

In principle, it is possible to derive sensitivity factors for photoelectron spectra from a theoretical model, involving photoionisation cross-sections,[16] electron mean free paths[4] and a number of assumptions concerning instrumental transmission functions,[15] sample roughness, etc. In practice, these assumptions may not be valid, and the theoretical prediction of photoelectron peak intensities usually shows a very poor agreement with experimental results.[15] The situation is worse in the case of Auger electron spectra, since the physical process of Auger emission is more complex than that of photoemission.[17]

In most cases, therefore, elemental sensitivity factors for both Auger and photoelectron spectra are derived experimentally, by recording the intensity of the relevant peak in a spectrum from a standard substrate (element or compound) under standard instrumental conditions.[18] Armed with these sensitivity factors, the surface elemental composition of an unknown sample may be determined using the following equation:

$$C_A = \frac{I_A/S_A}{\sum_n (I_n/S_n)} \times 100\% \tag{1}$$

where C_A is the concentration of element A, expressed as a percentage; I_A is the intensity of the peak from element A in the electron spectrum; and S_A is the experimentally derived sensitivity factor for element A.

Although the use of a single, experimentally derived, sensitivity factor for each element still involves a number of assumptions (e.g. that the form of the dependence of the electron mean free path on kinetic energy is independent of the nature of the sample),[4] this method of quantification can usually be relied upon to give elemental compositions to within 20% of the true value in most cases, and often to within 5% for well-characterised surfaces.

4.2. Intensity measurements

Whatever the procedure used for quantitative surface analysis by electron spectroscopy, it is necessary to make meaningful determination of the intensities of selected spectral features in the raw data. The intensity of an Auger or photoelectron feature in the electron spectrum is defined as the total number of Auger or photoelectrons contributing to that feature, $\int N(E)\, dE$, i.e. the total area under the feature (Fig. 7a). All such features, however, are superimposed on a non-uniform background of inelastically scattered electrons, and the main problem in intensity measurements is defining the shape of this background.[13] The problem is more acute in Auger electron spectroscopy, where broad and complex elemental features are superimposed on a strongly sloping inelastic background (Fig. 7b).

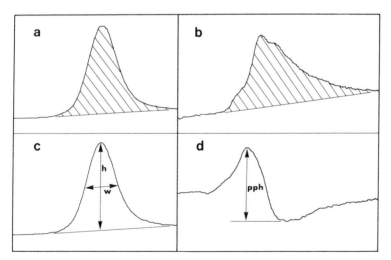

FIG. 7. Intensity measurements in electron spectroscopy. (a) Peak intensity (area) of a photoelectron feature. (b) Peak intensity (area) of an $N(E)$ Auger feature. (c) Approximation $(h.w)$ to the intensity of a photoelectron feature. (d) Approximation (peak-to-peak height, pph) to the intensity of a $dN(E)/dE$ Auger feature.

A rigorous intensity measurement can therefore be difficult, but it is possible to use a number of simplifications in order to make the analysis more convenient without increasing the probable errors significantly above the figures quoted in section 4.1.

The intensity of a photoelectron feature may be measured relatively easily, if on-line computing facilities are available, and a simple expression for the inelastic background is chosen (e.g. linear or stepped). However, a relatively close approximation to the peak area, for symmetric peaks at least, may be obtained by taking the product of the peak height and the peak width at half maximum ([$h.w$], Fig. 7c). Indeed, if it is assumed that the width of any particular feature is invariant between the standard and unknown, and between different unknowns, the peak height alone (h) may be used to give a measure of intensity.

Auger electron spectra are usually obtained as the first derivative of the electron current ($dN(E)/dE$, see section 2), and are thus not directly conducive to peak area measurement. However, a useful parameter that may be easily obtained from first-derivative spectra is the peak-to-peak height (pph) of the derivative (Fig. 7d). It is easily shown that, for Auger features of a similar shape and width, the pph is proportional to the peak area. The use of pph as a measure of Auger peak intensity is almost universal, and does not, in fact, involve any more assumptions than the use of peak height measurements in photoelectron spectroscopy.

In most cases, the procedures described above can yield relatively accurate results, and provide information about the sample surface that cannot be obtained by any other means. However, it is also possible to arrive at erroneous conclusions regarding the surface composition if sufficient data are not available.

The most common error arises from the fact that the electrons which contribute to the spectrum originate within the top 2 nm or so of the sample surface. Hence, the measured surface composition is, in fact, the *average* composition over a few atomic layers. It is, therefore, not possible to distinguish between, for example, a sample with a single atomic layer of an element X on the surface, and a sample with a certain amount of element X distributed evenly throughout the top 2 nm of surface. Such differences can be detected, however, if angular studies (section 5) or composition–depth profiling studies (section 6) are made.

Another aspect of surface structure, that of roughness, may also lead to anomalous interpretations of the quantified results of electron spectra. For any part of the sample surface to contribute to the observed electron spectrum, it is necessary for it to be in line of sight with both the excitation source and the electron energy analyser. Consideration of the geometry involved shows that the electron spectrum from microscopically rough samples will originate predominantly from the peaks of the surface topography, and the contribution from the troughs will decrease with increasing roughness, and with increasing angle between the excitation source and energy analyser.

Finally, the surface composition may have changed significantly during sample handling, mounting and introduction into the spectrometer, and even during the analysis itself. These changes, which may be due to contamination, desorption into the vacuum, or decomposition by the excitation source, are more apparent on high vapour pressure materials such as organic and polymeric samples. Methods of minimising these changes are discussed in the following section.

5. ANALYTICAL PROCEDURES OF SURFACE ANALYSIS

The purpose of this section is to describe the basic procedures necessary to obtain the desired information about the surface composition of solid samples.

5.1. The sample
Carbon is ubiquitous and even a supposedly 'clean' sample surface will show a significant carbon contribution to the electron spectrum, due to the presence of one or more layers of adsorbed hydrocarbons and oxides of carbon. It is possible to minimise the amount of contamination to which a surface is subjected, and this is desirable for the following reasons:

(a) Contamination is often an important factor in determining, for instance, the adhesive properties of surfaces.[19,20] Contamination arising from careless sample handling during, and just prior to, surface analysis may, therefore, give rise to confusing or conflicting results.

(b) A contamination layer will attenuate the electron signal from the underlying surface, and important features in the spectrum may be masked in this way. Although the contamination layer may be removed by *in situ* ion etching, it is preferable to avoid contamination in the first place.

(c) Some forms of contamination (e.g. lactic acid in perspiration) can react with the sample surface, and thereby alter the surface composition significantly.

It is generally not advisable to attempt to remove contamination before sample introduction into the electron spectrometer (by, for example, wiping the sample surface or washing in a solvent), since such procedures almost invariably alter the surface composition of the sample itself. Even attempting to remove particulate contamination with the aid of a proprietary aerosol dust remover can leave a monolayer of adsorbed fluorocarbons on the sample.

If the sample is handled carefully (with tweezers wherever possible), mounted in a suitable manner (see section 3.2) and analysed as soon as possible after the surface has been created, it should be possible to reduce contamination to a minimal level. Even if it is possible to create the surface for analysis inside the spectrometer itself (as with, for example, *in situ* fracture studies), contaminants from the residual vacuum (hydrocarbons in diffusion pumped systems, water and oxides of carbon and nitrogen in all systems) may still interfere.[21] In general, samples can be expected to remain 'clean' (where 'clean' means less than 10% of a monolayer of contaminants) for a period of at *least* 1000s in vacua of 10^{-10} Torr or less.

In fact, the problem of desorption of species from the sample surface into the vacuum (sometimes accelerated by the influence of the excitation source) is much more acute than that of contamination by the residual vacuum. Again, rapid analysis minimises this effect.

If necessary, contamination can be removed by ion etching just prior to or during analysis, but the ion beam can itself induce compositional changes in the sample surface[22] (and generally leads to a loss of chemical information in photoelectron spectroscopy), and so must be used with care.

Where it is necessary to prepare a clean surface for the purpose of *in situ* experimentation, this may be done in a number of ways. Metal surfaces can often be cleaned by heating to a high temperature for repeated short periods ('flashing'). Flashing may be used in conjunction with gas–surface reactions in order to remove contaminants (e.g. oxygen to remove carbon; hydrogen to remove oxygen).[23] Fracturing *in situ* can also be used to prepare a clean surface of most materials, and ion etching can be used on almost any material.

5.2. Auger electron spectroscopy
AES is often used in preference to XPS because of the ability to carry out

'point' analyses (with a lateral resolution of $0.5\,\mu m$) with the former technique (see section 3.1). Thus, not only may any particular area on the surface be selectively analysed, but it is also possible to produce chemical maps in AES, which show the distribution of any particular element over the whole surface. Such facilities are important in, for example, studies of adhesive failure,[24] of fracture surfaces (where grain boundary segregation may be of interest) or of corrosion products (where the distribution of elements may give some insight into the mechanisms involved). Because of the possibilities of electron beam damage on some surfaces, however, it is preferable to carry out analyses with large spot diameters ($> 100\,\mu m$), or defocused beams (thereby reducing the beam current density), whenever spatial information is not required.

In general, AES provides elemental information (i.e. an elemental analysis) only. The Auger peaks of many elements, however, show significant changes in position or fine structure in different chemical environments.[25] Thus, for example, it is possible to distinguish between elemental and oxidised aluminium and silicon, and between organic carbon and carbide-type carbon, in AES (Fig. 8).

The drawbacks of AES centre around the high current densities employed in the excitation source (usually several $A\,cm^{-2}$ at high resolution), which results in charging problems on rough, insulating samples, and beam damage problems on sensitive surfaces. These artefacts can be minimised by working at low beam energies and currents as often as possible, and the problems are probably overemphasised in the literature.

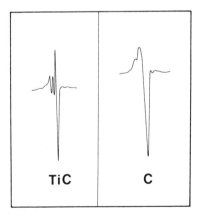

FIG. 8. The carbon Auger peaks from titanium carbide and carbon, illustrating the chemical bonding information obtainable from AES.

5.3. Photoelectron spectroscopy

XPS is often used in preference to AES when delicate surfaces, such as organic molecules,[26] polymers[27] and other high vapour pressure solids are to be studied. This is because beam damage is rare in XPS; not only is the X-ray source spread over a large area (usually about $1\,cm^2$), but the total power of the beam is much lower than that of the conventional electron sources used in AES. For the same reason, sample charging is not a significant problem in XPS, although it does still occur to a small extent on insulators.

As with AES, the XPS spectrum can be interpreted to give a quantitative elemental analysis of a sample surface. In addition, because XPS features tend to be sharper and less complicated than AES features, it is possible to measure peak positions to a high degree of precision (0·1 eV), and thereby obtain information about the oxidation state or chemical environment of the detected elements. Examples include the differentiation of oxides from hydroxides,[28] sulphides from sulphates[29] and carbides from organic carbon from carbonates,[22,30] as well as the determination of the oxidation state of most metals (Fig. 9). Unfortunately, much of this chemical information is destroyed by ion beam etching (carbonates may be decomposed to oxides,[22] and oxides reduced to the metal), so care must be exercised when removing contamination, or composition–depth profiling, with the aid of ion etching. These problems may be circumvented, to some extent, by the use of angle-resolved spectra. Briefly, at higher (grazing) emission angles, the electron spectrum becomes, effectively, more surface sensitive, since electrons from any particular depth are forced to travel through a greater amount of material. The exact relationship is given by

$$I_x = 1 - e^{-1/\lambda \cos \theta} \qquad (2)$$

where I_x is the fractional contribution from the surface layer, λ is the electron mean free path in monolayers, and θ is the electron emission angle (relative to the surface normal). Thus, by varying the emission angle from $0°$ to $70°$, it is possible to vary the fractional contribution from the surface layer from as little as 0·18 to as much as 0·44, and thereby obtain information about the depth variation of detected elements within the surface layers.[31]

The main disadvantage of XPS is the inability to focus X-rays, and hence the lack of spatial (lateral) resolution in the spectra. In most cases, photoelectron spectra represent the average surface composition over

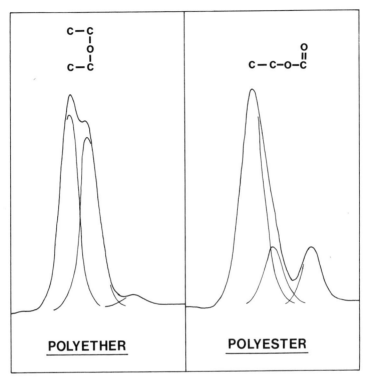

FIG. 9. The carbon photoelectron peaks from polyether and polyester samples, showing contributions from alkyl, ether and carboxyl species to the spectrum, illustrating the chemical bonding information obtainable from XPS.

several square millimetres, although spectrometers are available with analysis areas as low as 1 mm^2.

6. COMPOSITION–DEPTH PROFILING

The provision of accurate composition–depth profiles by AES or XPS is extremely important in the elemental and chemical characterisation of thin films, surface coatings and surface treatments. The technique is particularly important as a means of investigating the interfacial region between the surface film and its substrate. Composition–depth profiles can be obtained in two ways depending on the depth of interest. Sputter-depth profiling is normally used to depths of ∼2 μm and thereafter the profile is obtained by tapering the surface by some mechanical means.

6.1. Sputter-depth profiling

In AES, profiles are usually obtained by sequential ion beam sputtering and surface analysis. A hot-filament type of ion gun is used to erode the surface by bombardment with 0·5–5 keV inert gas ions (Ar^+ or Xe^+). This type of ion gun produces an essentially mono-energetic beam which can be focused into a spot, the ion current distribution is Gaussian and hence the beam develops a Gaussian-shaped crater in the sample surface which is typically 2 mm in diameter. The ion current densities achieved using this method are about $200\,\mu A\ cm^{-2}$. Alternatively, the ion beam can be raster-scanned, thereby producing a square profile in the surface. In both cases, the electron beam used for analysis is very much smaller than the crater, and the analysis can be performed at the bottom of the crater without taking any edge-effect contributions from the crater walls, as illustrated in Fig. 10. Since the sputtering yields of most materials are greater than 1 atom ion^{-1} for the ion energies used, etching rates of $25\,nm\ min^{-1}$ are typical.

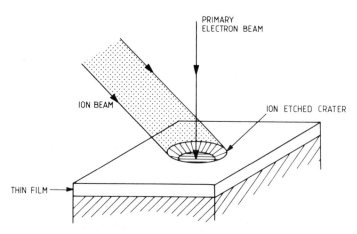

FIG. 10. A schematic diagram illustrating the use of a static ion beam to obtain a composition–depth profile through a thin film by AES.

When the depth distribution of only a few specified elements is required, the Auger peak from each element can be selectively scanned during simultaneous ion bombardment, using a multiplexed control unit. The amplitude of the selected Auger peaks can be plotted directly on a point plotter as a function of sputtering time. This speeds up the analysis considerably, but care must be exercised since some materials, notably aluminium, produce a high intensity of ion-induced Auger electron emission which causes uncertainty in quantitative analysis.[32]

Sputter-depth profiling can be useful at two levels. On the finest scale, a depth profile through a depth of < 10 nm enables the surface and sub-surface composition to be compared. The shape of the profile can be used to determine, for example, if a surface impurity has arisen from a source extraneous to the material, or has segregated by some mechanism from the bulk. For thicknesses up to $2\,\mu$m, the technique can be used to characterise the composition of thin films or to investigate the interface between the thin film and its substrate. There is an enormous variety of applications of this kind. The technique has been applied to a wide range of protective oxides and thin films. An example of its use in the study of surface pretreatments is shown in Fig. 11, from the work of Solomon and Baun,[20] aimed at improving the structural adhesion of aluminium alloys using epoxide adhesives for the aircraft industry. The Auger spectrum taken from the as-received 6061 Al alloy surface is shown in Fig. 11(a) together with a composition–depth profile performed through the oxide layer. Surface pretreatment is necessary to remove this magnesium-rich layer and Figs. 11(b), (c) and (d) compare the composition of the surface following three different treatments. The results show that the three treatments produce markedly different oxide thicknesses. The thinnest oxide layer remains following the alkaline–acid treatment, while the bifluoride and chromate conversion coating treatments do not significantly alter the oxide.

In practice, the erosion of surfaces under ion bombardment is not uniform and this leads to the development of ion-induced surface roughness under the beam.[33] This mechanism, together with the effects of cascade mixing and radiation enhanced diffusion, leads to an inevitable deterioration of depth resolution with depths. As a guide, results obtained from a multilayer Ni/Cr thin film structure suggest that the depth resolution in metals deteriorates as a function of $Z^{1/2}$, where Z is the depth.[34] However, surface topography can occur in metals as a result of the erosion of ion-induced structural defects, and since this mechanism does not occur in insulators or in semiconductors whose surfaces are made amorphous by ion bombardment, the depth resolution of depth profiles obtained from insulators and semiconductors is usually better than that obtained from metals and alloys.[35] Although it has been shown recently that the simultaneous use of two ion guns can partially suppress the formation of ion-induced surface topography,[36] acceptable results are obtained only up to a depth of about $2\,\mu$m.

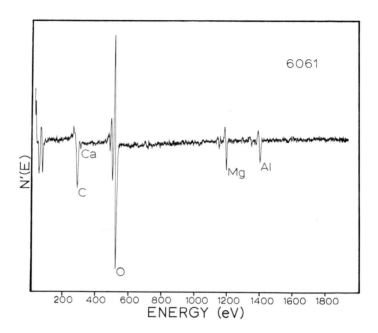

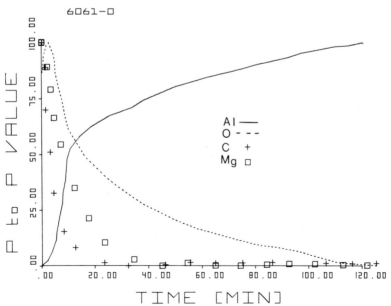

FIG. 11. The use of AES in a comparison of various surface pretreatments on 6061 A1 alloy sheet. In each case the Auger spectrum taken from the surface is shown (top) together with a sputter-depth profile through the oxide layer (bottom). (a) The as received sheet.

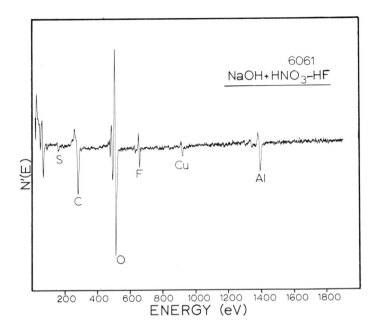

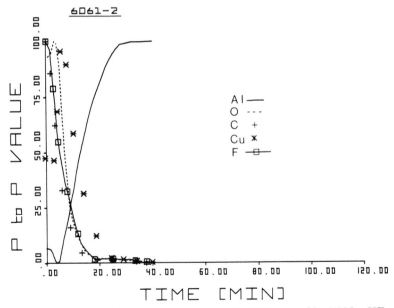

FIG. 11.—*contd.* (b) Surface treatment with NaOH followed by HNO_3–HF.

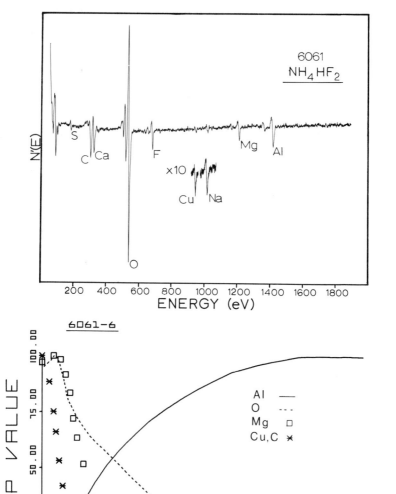

FIG. 11.—*contd.* (c) Surface treatment with NH_4HF_2.

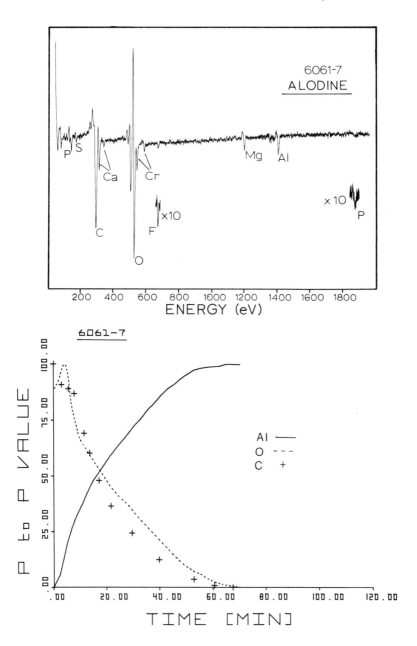

FIG. 11.—*contd.* (d) Surface treatment with Alodine (a commercial etch solution, a conversion mixture of phosphoric acid, chromic acid and NaF). From Solomon and Baun, ref. 20.

It is difficult to define accurately the depth scale of sputter-depth profiles, although fairly accurate estimates can be made by using standards. If the sputtering yield S of the material is known, the depth eroded is approximately given by

$$\text{Depth of erosion} = \frac{itS}{e} \cdot \frac{M}{N\rho}$$

where i = ion current density, t = sputtering time, S = sputtering yield, M = molecular weight, e = electronic charge, N = Avogadro's number and ρ = density.

Ion bombardment can also cause compositional artefacts to occur, due to the preferential sputtering, or by ion-induced chemical decomposition. The uncertainties that arise due to these mechanisms can be overcome to some extent by the use of standards close to the composition to be analysed and by incorporating the effects as a matrix correction factor in quantitative analysis.[37]

The combination of sputter-depth profiling and surface analysis by XPS is also extremely useful. A cold-cathode ion source is normally used, since a much broader surface area must be etched because of the lack of spatial resolution in XPS. Such ion sources produce a beam with a distribution of ion energies and a high proportion of neutrals and are generally less amenable to controlled sputter-erosion than the hot-filament sources used in AES. The advantages of XPS in composition–depth profiling are twofold. Firstly, analysis using XPS can be conducted on delicate materials including polymers, rubbers and organic coatings; secondly, careful use of the technique sometimes allows chemical information to be obtained as a function of depth.

However, since in XPS a comparatively large area (approximately $1\,\text{cm}^2$) must be etched, the problem of ion-induced surface roughness is more acute than in AES and the depth over which useful information may be obtained is more restricted.

6.2. Taper-sectioning techniques
Several methods have been used to develop a taper section for depth profiles $> 2\,\mu\text{m}$. Conventional angle lapping has been used with some success,[38–40] but the technique is laborious and is unsuitable for curved surfaces. Another technique which has been used on glass surfaces involves the use of external high-rate ion etching equipment to develop a ramp through the surface,[41] but this technique is slow and involves the incorporation of ion-induced artefacts discussed above. More recently, a

ball-cratering technique has been developed which seems to overcome many of these difficulties.[42,43] In this technique a lightly loaded rotating ball, coated with fine diamond paste, is used to fashion a well-defined spherical crater in the sample surface. The geometry of the crater is shown in Fig. 12. Since the radius of the ball is known, the diameter of a

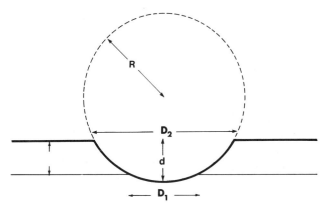

FIG. 12. Schematic diagram illustrating the geometry of a ball crater through a surface coating of thickness d which is given by $(D_2^2-D_1^2)/8R$.

crater needed for a given depth can be specified and measured. The appearance of a crater though an electrodeposited zinc coating (33 μm) is shown in Fig. 13. After sputter-cleaning, the composition–depth profile can be obtained either by point-by-point analysis down the crater wall or by using element line-scans across the crater. A composition–depth profile through the electroplated zinc coating in Fig. 13 is shown in Fig. 14. In addition to its use for characterising surface coatings and treatments, the technique also enables the interface to be examined in fine detail[43] and this is often crucial in determining why the coating does not adhere properly. Consider the image of the crater shown in Fig. 13. The interface between the coating and substrate is clearly defined by a sharp change of contrast. At points near the substrate, the coating is extremely thin and thus a sputter-depth profile at the point A, for example, locates the interface within 100 nm or so.

    The ball cratering technique can be used to obtain composition–depth profiles in the range 1–100 μm. It is ideal for the investigation of a whole range of industrially important protective coatings such as those produced by electrodeposition, hot-dipping and spraying, and a range of surface treatments such as nitrocarburising, aluminising and boriding.

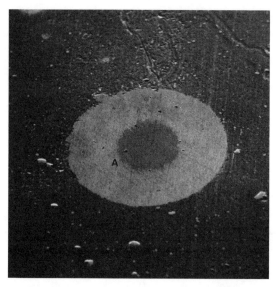

FIG. 13. Appearance of a ball crater through an electrodeposited zinc coating on mild steel. Point A corresponds to a position suitable for sputter-depth profiling of the coating–substrate interface. From Walls, Hall and Sykes, ref. 43.

Other potential areas of application include the analysis of thick oxide layers, multilayer coatings, and all forms of welding and bonding.[43]

7. SUMMARY

In this chapter we have discussed the principles and applications of Auger electron spectroscopy and X-ray photoelectron spectroscopy. Both techniques have been shown to be powerful methods of surface analysis combining surface sensitivity with an amenable range of detection sensitivity through the elements with atomic number $Z > 2$. The sample can take virtually any physical shape or form and needs no special preparation prior to analysis. However, before the techniques may be applied to a particular system, it is necessary to appreciate fully their particular advantages and limitations.

The chief advantage of Auger electron spectroscopy is its high spatial resolution. This makes the technique a viable tool for the identification and analysis of small areas, and also the best suited to composition–

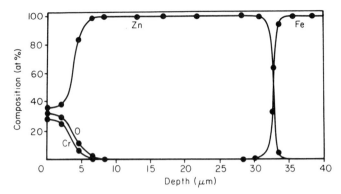

FIG. 14. Composition–depth profile through the electroplated zinc coating using ball cratering and AES. From Walls, Hall and Sykes, ref. 43.

depth profiling. It can be used on a range of materials including insulators, although charging problems can sometimes occur on rough insulating surfaces. A further problem with insulators sometimes occurs because the electron beam creates potential gradients in the surface which can cause the diffusion of ionic species (e.g. Na^+ or F^-).[44] A more serious problem occurs on delicate materials such as organic coatings and polymer surfaces, where the electron beam can cause severe damage by local heating at the point of analysis. In the study of adhesion and surface pretreatment, it is likely that AES will find use mainly in the study of metals and alloys and in those circumstances where spatial resolution is crucial. Examples of work carried out using AES in this field include the characterisation of anodised aluminium alloy adherends,[45] the locus of failure of adhesive joints,[24] the nature of polysiloxane/metal interfaces,[46] and the effects of surface contamination on adhesive bonding.[20]

The use of soft X-rays does not degrade or alter the composition of the most delicate surfaces, and for this reason XPS is amenable to a much wider range of materials. A further advantage of this technique is its ability to provide chemical information from the shifts in binding energy. Its main limitation is the lack of spatial resolution, which means that the technique can provide only an average surface analysis over areas > 1 mm^2, and is less suitable for composition–depth profiling. However, because XPS is well suited for use with polymers, it has been particularly successful in the study of polymer to metal adhesion. For example, the technique has been used to determine the nature and extent of modification of various polymer surfaces, following treatment with ozone,[47] flame treatment,[48] chromic acid etching,[49] sodium in liquid ammonia etching,[50] radiation exposure[51] and ion beam etching.[52] It has also been applied to the characterisation of the polymer–metal interaction in systems such as polyethylene–aluminium,[47] poly(vinyl alcohol)–copper,[53] polystyrene with various metals[54,55] and rubber to brass.[56] More fundamental studies, such as on the bonding of various organic monolayers on oxide surfaces,[26] adhesion failure in polymer–metal systems,[29] and the investigation of refractory coating adhesion on steel,[57] have also been carried out with the aid of XPS.

Both techniques of electron spectroscopy, AES and XPS, have developed considerably in recent years. Undoubtedly, their use in the study of surface pretreatments, adhesives and various bonding systems will grow as their potential uses and applications are fully realised.

REFERENCES

1. Harris, L. A., *J. Appl. Phys.*, **39** (1968), 1419.
2. Davis, L. E., MacDonald, N. C., Palmberg, P. W., Riach, G. E. and Weber, R. E., *Handbook of Auger electron spectroscopy*, 2nd ed., Physical Electronics Industries, Minnesota 1976.
3. McGuire, G. E., *Auger electron spectroscopy reference manual*, Plenum, New York 1979.
4. Penn, D. R., *Phys. Rev. B.*, **13** (1976), 5248.
5. Seah, M. P. and Dench, W. A., *Surf. Interface Anal.*, **1** (1979), 2.
6. Williams, P. M., in *Handbook of X-ray and ultraviolet photoelectron spectroscopy*, D. Briggs (Ed.), Heyden, London 1977, p. 313.
7. Krause, M. O., *Chem. Phys. Lett.*, **10** (1971), 65.
8. Kunz, C., in *Topics in applied physics*, Vol. 27, L. Ley and M. Cordona (Eds.), Springer-Verlag, Berlin 1979, p. 299.
9. Thompson, M., *Talenta*, **24** (1977), 399.
10. Avanzino, S. C. and Jolly, W. L., *J. Amer. Chem. Soc.*, **100** (1978), 2228.
11. Brundle, C. R., Roberts, M. W., Latham, D. and Yates, K., *J. Electron. Spectrosc. Relat. Phenom.*, **3** (1974), 241.
12. Palmberg, P. W., Bohn, G. K. and Tracy, J. C., *Appl. Phys. Lett.*, **15** (1969), 254.
13. Read, F. H., Comer, J., Imhof, R. E., Brunt, J. N. H. and Harting, E., *J. Electron. Spectrosc. Relat. Phenom.*, **5** (1974), 643.
14. Powell, C. J. and Larson, P. E., *Appl. Surf. Sci.*, **1** (1978), 186.
15. Seah, M. P., *Surf. Interface Anal.*, **2** (1980), 222.
16. Scofield, J. H., *J. Electron. Spectrosc. Relat. Phenom.*, **8** (1976), 129.
17. Palmberg, P. W., *Anal. Chem.*, **45** (1973), 549A.
18. Evans, S., Pritchard, R. G. and Thomas, J. M., *J. Phys. C: Sol. State Phys.*, **10** (1977), 2483.

19. Dwight, D. W. and Wightman, J. P., in *Surface contamination*, Vol. 2., K. L. Mittal (Ed.), Plenum Press, London 1979, p. 569.
20. Solomon, J. S. and Baun, W. L., in *Surface contamination*, Vol. 2, K. L. Mittal (Ed.), Plenum Press, London 1979, p. 609.
21. Bryson, C. E., Scharpen, L. H. and Zajicek, P. L., in *Surface contamination*, Vol. 2, K. L. Mittal (Ed.), Plenum Press, London 1979, p. 687.
22. Christie, A. B., Sutherland, I. and Walls, J. M., *Vacuum*, in press.
23. Barrie, A., in *Handbook of X-ray and ultraviolet photoelectron spectroscopy*, D. Briggs (Ed.), Heyden, London 1977.
24. Gettings, M., Baker, F. S. and Kinloch, A. J., *J. Appl. Polym. Sci.*, **21** (1977), 2375.
25. Haque, C. A. and Spiegler, A. K., *Appl. Surf. Sci.*, **4** (1980), 214.
26. Anderson, H. R. and Swalen, J. D., *J. Adhes.*, **9** (1978), 197.
27. Holm, R. and Storp, S., *Surf. Interface Anal.*, **2** (1980), 96.
28. Servais, J. P., Lempereur, J., Renard, L. and Leroy, V., *Brit. Corros. J.*, **14** (1979) 126.
29. Van Ooij, W. J., Kleinhessenlink, A. and Leyenaar, S. R., *Surf. Sci.*, **89** (1979), 165.
30. Brainard, W. A. and Wheeler, D. R., *J. Vac. Sci. Technol.*, **15** (1978), 1800.
31. Fadley, C. S., *Progr. Sol. State Chem.*, **11** (1976), 265.
32. Powell, R. A., *J. Vac. Sci. Technol.*, **15** (1978), 125.
33. Smith, R. and Walls, J. M., *Surf. Sci.*, **80** (1979), 557.
34. Hofmann, S., *Talanta*, **26** (1979), 665.
35. Webber, R. D. and Walls, J. M., *Thin Solid Films*, **57** (1979), 201.
36. Sykes, D. E., Hall, D. D., Thurstans, R. E. and Walls. J. M., *Appl. Surf. Sci.*, **5** (1980), 103.
37. Hall, P. M. and Morabito, J. M., *Surf. Sci.*, **83** (1979), 391.
38. Tarng, M. L. and Fisher, D. G., *J. Vac. Sci. Technol.*, **15** (1978), 50.
39. Moulder, J. F., Jean, D. G. and Johnson, W. C., *Thin Solid Films*, **64** (1979), 427.
40. Chubb, J. P., Billingham, J., Hall, D. D. and Walls, J. M., *Metals Technol.*, **7** (1980), 293.
41. Chappell, R. A. and Stoddart, C. T. H., *J. Mat. Sci.*, **12** (1977), 2001.
42. Thompson, V., Hintermann, H. E. and Chollet, L., *Surf. Technol.*, **8** (1979), 421.
43. Walls, J. M., Hall, D. D. and Sykes, D. E., *Surf. Interface Anal.*, **1** (1979), 204.
44. Chappell, R. A. and Stoddart, C. T. H., *Phys. Chem. Glasses*, **15** (1974), 130.
45. Solomon, J. S. and Hanlin, D. E., *Appl. Surf. Sci.*, **4** (1980), 307.
46. Gettings, M. and Kinloch, A. J., *Surf. Interface Anal.*, **1** (1979), 189.
47. Briggs, D., Brewis, D. M. and Konieczko, M. B., *Eur. Polym. J.*, **14** (1977), 1.
48. Briggs, D., Brewis, D. M. and Konieczko, M. B., *J. Mat. Sci.*, **14** (1979), 1344.
49. Briggs, D., Zichy, V. J. I., Brewis, D. M., Comym, J., Dahm, R. H., Green, M. A. and Konieczko, M. B., *Surf. Interface Anal.*, **2** (1980), 107.
50. Von Brecht, H., Mayer, F. and Binder, H., *Angew. Makromol. Chem.*, **33** (1973), 89.
51. Yamakawa, S., *Macromol.*, **12** (1979), 1222.
52. Sovey, J. S., *J. Vac. Sci. Technol.*, **16** (1979), 813.
53. Burkstrand, J. M., *J. Vac. Sci. Technol.*, **16** (1979), 363.

54. Burkstrand, J. M., *J. Vac. Sci. Technol.*, **16** (1979), 1072.
55. Burkstrand, J. M., *Appl. Phys. Lett.*, **33** (1978), 387.
56. Van Ooij, W. J., *Rubber Chem. Technol.*, **52** (1979), 605.
57. Brainard, W. A. and Wheeler, D. R., *J. Vac. Sci. Technol.*, **16** (1979), 31.

Chapter 3

ANALYSIS OF METAL SURFACES BY ION SPECTROSCOPY

W. L. BAUN
Wright Patterson Air Force Base, Ohio, USA

1. INTRODUCTION

There is an increasing need for sensitive analytical techniques to examine the changes caused by surface treatments as outlined in Chapter 1. Empirical bondability methods such as the water break test may indicate wettability and subsequent bondability, but they tell us little of long time durability of the bond which may depend on surface chemistry. A number of analytical techniques have been developed for characterising solid surfaces. Some of these techniques, using electrons and photons as probes of the surface chemistry, have been described in this volume by other authors. In this chapter, methods of surface analysis using beams of ions will be described.

2. PRINCIPLES AND PROCEDURES

2.1. Ion scattering spectrometry (ISS)

Use of low energy backscattered ions to characterise a surface is a relatively recent development. The method has been reviewed by Buck.[1] High energy ions had been used in the past to analyse surfaces, but it was not until Smith[2] used low energy (1 KeV) noble gas ions to probe the surface of a variety of materials that the technique came into popular use. It was found, from this and other work, that when the energy of ions was lowered, the scattered ion spectra became simpler and sharper and approached the behaviour expected on the basis of a binary scattering event from a single surface atom. Therefore, the energy E_1 retained by an ion of mass M_{ion} with an incident energy E_0 after scattering from an atom of mass M_{atom} through an angle θ is given by eqn. (1), which is based on the conservation of kinetic energy and momentum (here M_{ion} is smaller than M_{atom}).

$$\frac{E_1}{E_0} = \frac{M_{ion}^2}{(M_{ion} + M_{atom})^2} \left\{ \cos\theta + \left(\frac{M_{atom}^2}{M_{ion}^2} - \sin^2\theta \right)^{1/2} \right\}^2 \tag{1}$$

For 90° scattering which is frequently used this reduces to a very simple relationship:

$$E_1/E_0 = (M_{atom} - M_{ion})/(M_{atom} + M_{ion}) \tag{2}$$

The experimental set-up for 90° scattering instruments is shown in Fig. 1(a) and is representative of original commercial instrumentation (3M Company, St Paul, Minnesota), which used a 127° electrostatic analyser. Recently, a great improvement in sensitivity was gained by the development of a cylindrical mirror analyser (CMA), substituted for the original electrostatic sector as shown in Fig. 1(b). The geometry of the CMA results in a scattering angle of 138°. The major advantage of low energy ion scattering is the extremely fine surface selectivity when low energy ions collide with the surface atom. The probability for neutralisation is very high because of the long residence time (10^{-14}–10^{-16} s). Only about one in 10^3 of the scattered particles retains a positive charge even after one collision. Therefore, the probability that an ion is still in the charged state after two or more collisions is very small. Since the detector responds only to charged particles, contributions from particles which scatter more than once are almost certainly neutral and not counted by an ion detector. These facts suggest that an instrument using time-of-flight methods to detect either ions or neutrals would be extremely useful. Such an instrument is currently in use.[3]

An inherent feature of ion scattering which may be considered an advantage or a disadvantage is the simultaneous sputtering of the surface as energy is transferred to the surface atoms from the ion beam. It is an advantage in that the concentration of the various atomic species may be followed with depth. On the other hand, it is a disadvantage because damage is being produced by the sputtering. Once the atom sputters from the surface the sample is changed, and an exact experiment on that spot may not be repeated. One positive feature of ion scattering compared with most other spectroscopy techniques is the simplicity of

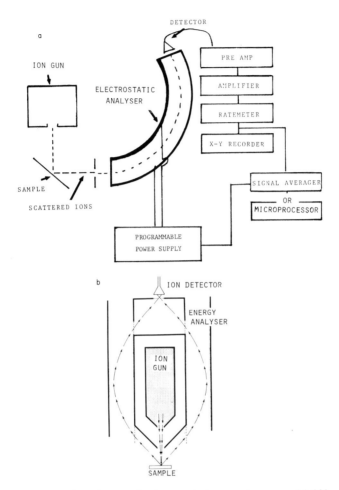

FIG. 1. Experimental equipment for ion scattering spectroscopy. (a) 90° scattering set-up representative of original commercial equipment; (b) cylindrical mirror analyser (CMA) representative of recent commercial equipment.

the spectra. Binary ion scattering results in one peak for each isotope of an element present. For instance, as seen in Fig. 2, the scattering of helium from aluminium oxide results in the appearance of only two peaks in the spectrum, one for oxygen and one for aluminium. Each peak is very sensitive to the amount present, but absolute quantitative analyses can be difficult since the scattered yield depends on scattering cross-

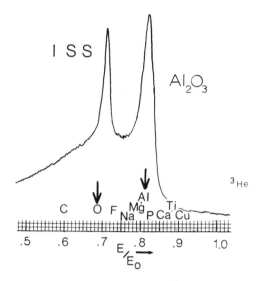

FIG. 2. Ion scattering spectrum of $^3He^+$ from Al_2O_3.

section and neutralisation efficiency, neither of which is well known for most elements. During the ion scattering experiment atoms are sputtered from the surface, allowing depth profiling analysis by the removal of the surface layers by the probe ion during the analysis. Use of helium ions gives a very slow rate of surface removal and while neon and argon provide much higher sputtering rates, the ion beam may be focused and rastered on the surface to reduce sputtering, while the signal is gated from the centre of the crater to reduce crater edge effects. The signal may be collected from the surface using the rastered beam to give a lateral analysis of the surface. Therefore, ion scattering provides a combination of in-depth analyses and lateral analyses to give a three-dimensional picture of the chemical makeup of the surface with depth.

2.2. Secondary ion mass spectrometry (SIMS)

When a low energy ion strikes the solid surface, it undergoes and produces a number of interactions which are illustrated in Fig. 3.[4] The process under consideration here is process 5 as seen in this figure (the reflected ion giving energy to the surface atom which is sputtered). The sputtered species which are removed from the surface are made up of both positive and negative ions as well as neutral particles. Neutral particles have much more abundance than ionic species and have also

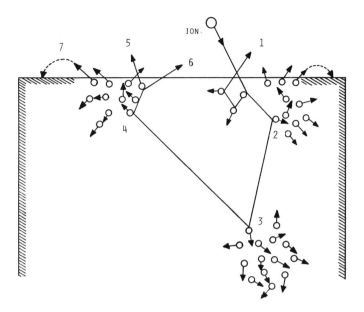

FIG. 3. Interaction of low energy ions with solid surfaces.[4]

been used for surface analysis. Surface analysis by SIMS falls into two categories, low current density sputtering and high current density sputtering. Categories are determined by the characteristics of the primary ion beam. A low current density sputtering analysis results in a very small fraction of the surface being disturbed, a result that approaches a basic requirement of a true surface analysis method. This is generally known as the Static SIMS (SSIMS) method. High current density sputtering removes more material and is required for preparing elemental depth profiles. In the high current density method, changes are seen in the surface and near surface regions. Table 1 shows the main features of SIMS as a surface analysis method.[5] Of the positive attributes listed, probably the extremely high sensitivity for many elements is the greatest advantage of SIMS. On the other hand, the large difference in sensitivity for different surface structures and chemical combinations is the largest negative factor involved in SIMS analyses. Isotopic identification and the sensitivity to hydrogen are two other important uses of the SIMS method. These advantages and disadvantages are reviewed by Werner[6] and quantitative aspects are discussed by the same author.[7]

Equipment for SIMS may be as simple as that shown in Figure 4(a), or

TABLE 1

MAIN FEATURES OF SIMS AS A SURFACE ANALYSIS METHOD[5]

Positive	Information depth in the 'monolayer range'
	Detection of all elements including hydrogen
	Detection of chemical compounds
	'Lateral resolution' in the range of atomic distances
	Isotope separation
	Extremely high sensitivity for many elements and compounds ($<10^{-6}$ monolayers)
	Quantitative analysis after calibration
	Negligible destruction of the surface (SSIMS)
	Elemental profiling (Dynamic SIMS)
Negative	Large differences in sensitivity for different 'surface structures' (factor of 1000)
	Problems in quantitative interpretation of molecular spectra
	Ion induced surface reactions

as complex as the ion microprobe schematic diagram shown in Figure 4(b). In a simple system a SIMS experiment requires a vacuum chamber to house the experiment, a sample holder, an ion source, an energy analyser and a mass analyser. In such simple systems the noble or reactive gas fills the system and the entire chamber including the ion gun and sample area are at approximately 1×10^{-5}–5×10^{-5} Torr. A more complicated type of instrument is one in which the performance is improved through the use of a differentially pumped vacuum system to produce ultra-high vacuum in the vicinity of the sample. This also allows the entry of a reactive gas in the sample chamber area while sputtering with a noble ion for studying chemical changes or reactions on the surface. Still another improvement, adding complexity, may be made to the SIMS instrument by the mass analysis of the primary beam. The energy filter is generally made up of several elements whose function is to optimise collection of the secondary ions and to filter and focus the ions at the entrance to the mass analyser. The mass analyser in simple systems is usually a quadrupole filter. More complex instruments, particularly in imaging equipment such as ion microprobes and the double focusing magnetic spectrometer, are frequently used, such as that shown in Figure 4(b).

It is recognised that SIMS has been used successfully as a stand-alone technique to solve many surface problems. However, it appears that the area of greatest use of SIMS is as a complement to other surface characterisation methods. The extremely high sensitivity for some ele-

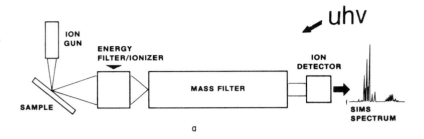

a

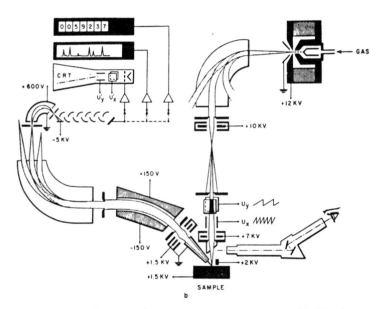

b

FIG. 4. Equipment for secondary ion mass spectrometry. (a) Simple system showing components necessary in ultra-high vacuum; (b) ion microprobe.

ments can be taken advantage of by using SIMS with other techniques in which these elements do not show high sensitivity. The SIMS technique is also ideal to use with fundamentally low resolution methods such as ion scattering. To separate and identify the adjacent masses which may be present at the sample surface, the most popular combination of instruments used so far has been ISS–SIMS and AES–SIMS. SIMS has also been used on scanning electron microscopes, allowing high quality imaging along with lateral and depth analysis of the sample.

2.3. Other techniques using an ion beam

2.3.1. Rutherford backscattering

When a surface is bombarded with a beam of ions in the MeV energy range, a small fraction of the incoming particles undergoes a Rutherford collision and is backscattered. A recent book has been published on this subject.[8] The experiment is conceptually simple but requires complicated and expensive equipment to carry it out. A typical experimental set-up for routine backscattering analysis is shown in Fig. 5.[8] Charged particles

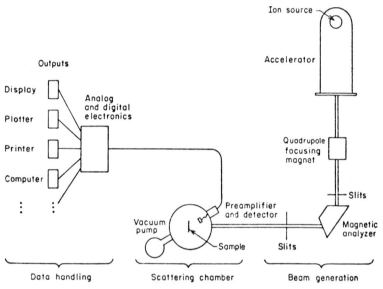

FIG. 5. Experimental equipment arrangement for ion backscattering analysis.[8]

are generated in an ion source, then the energy is raised to several MeV by an accelerator, usually a van de Graaff. The high energy beam passes through a series of devices which will focus the beam and filter it for a particular type of particle and energy. The beam then enters the scattering chamber and impinges on the sample. Some of the backscattered particles strike the detector where they generate an electrical signal. This signal is amplified and processed with analogue and digital electronics. The energy E of the ion scattered through an angle of 180° is related to the initial energy E_0 by the equation:

$$E = \left(\frac{M-m}{M+m}\right)^2 E_0 \qquad (3)$$

where m is the mass of the scattered particle and M is the mass of the target atom. Thus energy analysis of the primary ions backscattered from a sample provides a mass analysis over a sample volume with depth. The relative sensitivity of different surface atoms is determined by the Rutherford cross-section for the element. There is also an effect of the sample thickness on the energy of the ion beam; therefore the energy spectrum is a convolution of a mass scale established by the backscattering process and a depth scale established by an energy loss of the ion beam before and after the backscattering event.

2.3.2. Surface composition by analysis of neutral ion impact radiation (SCANIIR)
In addition to causing sputtering at the surface, a primary ion beam causes the emission of photons in the visible and ultraviolet region.[9] By analysing the light output with an optical spectrometer, the chemical analysis of the surface and depth profiling may be obtained. This technique, SCANIIR, has not come into general use but would appear to be useful for surface analysis, especially as a complement to some of the other methods using ion beams.

3. INFORMATION OBTAINED BY ION SPECTROSCOPY

The two ion beam techniques most popularly used for true surface analysis are ion scattering spectrometry and secondary ion mass spectrometry. Both of these techniques have certain capabilities and limitations. These features are summarised in Table 2.[10]

These two techniques may be applied in a gentle manner using a near-static ion beam to produce little change in the surface, and also in a mode in which chemical profiling with depth is possible. ISS can detect all of the elements heavier then helium in the periodic table. The sensitivity variation across the periodic table is probably less than one order of magnitude. SIMS provides a distinct advantage of being able to analyse, in principle, all of the elements in the periodic table. Being able to identify isotopes is a definite advantage. The sensitivity of the SIMS technique can vary by several orders of magnitude (perhaps up to four) due to a rapidly changing secondary ion yield caused by matrix and chemical effects.[7] The ability of ion scattering to resolve different elements in a complete unknown is at times somewhat limited. There are few intrinsic limitations or spectral interferences, but the technique is

TABLE 2
COMPARISON OF ION SPECTROSCOPY METHODS

Parameter	Ion scattering spectrometry (ISS)	Secondary ion mass spectrometry (SIMS)
Principle	Elastic binary collision with surface ion	Sputtering of surface atoms by ion beam
Probe	∼ 1 to 3 keV ions	∼ 1 to 3 keV ions
Signal	Ion current versus energy	Ion current versus mass
Applicable elements	$Z \geq 3$	All (if positive and negative SIMS)
Surface sensitivity	High	Variable
Elemental profiling	Yes	Yes
Image-spatial analysis	Yes	Yes
Spectral shift	Possible, but generally no	No
Information on chemical combination	Yes in fine features, but generally no	In some cases (fingerprint spectra)
Quantitative analysis	Yes	Maybe with similar standards
Influence of operating conditions and matrix	No	Yes
Isotopic analysis	Yes, in principle, but generally no because of resolution limits	Yes
Beam induced surface changes	Yes, sputtering damage	Yes, sputtering damage

fundamentally a low resolution technique in which there is some uncertainty as to the exact identity of a given line. Specificity may be improved by going to a scattering ion closer to the mass of the unknown element; that is, we should use helium for the light elements, neon for intermediate mass elements and argon for the heavy elements.

Ion scattering gives very little information on the chemical combination of the element detected in the sample; however, recently discovered yield variations[11] and the use of other fine features in the spectrum[12] give some possibility of using ion scattering to determine chemical species at the surface. The appearance of cluster ions in the secondary ion mass spectrum gives a good possibility of using SIMS to determine chemical combinations.[5] The interpretation of such spectra is extremely complicated and has to be treated with a great deal of care. Molecular ions can be dislodged from the surface and give some idea of the chemical combination, but molecular complexes may also be synthesised in the gas phase above the sample surface. The presence of such ions in the mass spectrum does not unequivocally prove the presence of such species in or on the sample itself.

4. APPLICATION OF ISS AND SIMS

4.1. Changes in chemistry due to surface treatments
Many chemical etching and oxidising treatments are used on metal and alloys to enhance a variety of properties. Numerous thermal pretreatments following fabrication improve strength, ductility, toughness or some other property. Each of these chemical or thermal treatments affects the composition of the surface either by introducing impurities, or by increasing or decreasing the concentration of alloying elements at the surface. Many impurities are present in the raw materials or are introduced during materials processing. Final fabrication of the materials, into the desired shape for adhesive bonding may also introduce contaminants. Prebonding treatments generally are designed to alter mechanically or chemically the original surfaces to enhance bondability. However, these treatments frequently leave behind contaminants which are deleterious to adhesive bonding. Finally, the environment and careless handling before and during the actual bonding can introduce contaminants or even physically alter the adherend surfaces. A summary of the sources of contaminants which could affect adhesive bonding and bond properties is shown in Table 3.[13]

TABLE 3
SOURCES OF CONTAMINANTS
OR SPECIES WHICH COULD
AFFECT ADHESIVE BONDING
AND BOND PROPERTIES[13]

1. Raw Materials Processing
   (a) Adherend
   (b) Adhesive
2. Prebonding Treatments
   (a) Chemical solution contributions
   (b) Alloy constituents
3. Environment
   (a) Storage and handling
   (b) Bonding

Very often the mill scale that remains on a material following final processing or rolling has little relation to the concentration of each element in the bulk. The ISS and SIMS spectra from such an initial 2024 aluminium alloy, as shown in Fig. 6, indicate high magnesium concentration at the surface. Conventional alkaline cleaning treatments do not etch the surface appreciably, leaving the surface magnesium rich. Such a surface, when adhesively bonded, may exhibit good initial bondability but poor long time durability when compared with bonded structures in which formation of aluminium oxide has been assured.

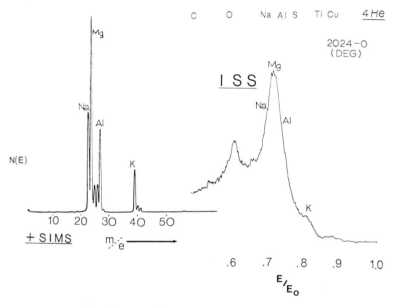

FIG. 6. ISS and SIMS data for 2024 Al alloy.

In addition to alloying elements adhering to the surface, the same phenomenon can occur with impurities. Table 4 lists some of the impurities that either exist naturally in the original ore, or are introduced during processing, heat treatment and fabrication or in the final processing.[13] In several areas of processing, such as the use of a flux in skimming of the molten bath, it appears that little thought is given to the possible consequences of adding such materials as sodium, potassium and chlorine to the melt. These materials, although existing perhaps only in the parts-per-million range in the bulk alloy, may diffuse to the surface to

TABLE 4

POSSIBLE SOURCES OF IMPURITIES IN ADHESIVELY BONDED ALUMINIUM ALLOYS

Process	Common Impurities	Source
(A) Al alloys		
Melting	Si ore	$Al + SiO_2$
	Fe impurity	$Al + Fe_2O_3$
Skimming	Na, K, Cl	Flux
Degassing	Cl, Mg, Zn	Flux
Remove inclusion	Na, Cl, K	Flux
Forming	C	Lubrication
Fabrication	C, S, Cl	Lubrication
	(soaps, oils, greases)	
Heat treating		
Solution heat treatment,	KNO_3, $KNO_3/NaNO_3$	Salt bath
500–550°C	$K_2Cr_2O_7$	Salt bath corrosion inhibitor
	Silica sand	Fluidised beds
Precipitation heat	H_2O, O, N, CO_2, CO	Furnace atmosphere
treatments,	SO_2, SO_3, NH, H_2	
150–250°C	Mg and others	Excessive temperature
	(diffusion of alloying elements through Al clad)	
(B) Epoxy adhesives		
Catalysis	Na, K	Sodium hydroxide
Prevent emulsification during washing	Na, S	Na_2HS
Polymerisation (special)	Li	LiOH
Intermediate production	B, F	Lewis acid BF_3
Dehydrohalogenation	Na, Al	Sodium aluminate
Synthesis of ester	Na, K, Cl	Na or K salt of acid or acid chloride
Epoxidisation	Cl	HClO
		C_4H_9ClO

provide virtually 100% coverage on an adherend surface. This interphase region may have entirely different wetting and water vapour transport characteristics, for instance, from the original prepared alloy.

After proper removal of the mill scale, the new oxide that is formed, often shows some unusual chemical composition. Such an example is shown in the elemental profile of the oxide formed on 2024-T3 aluminium alloy as etched with the commonly used FPL treatment seen in

Fig. 7. In this study it was found that this etch and other similar etches produced different oxide thicknesses and different widths of the copper profile depending on the time and temperature of the FPL etch.[14] This work, which involved a combination of ISS, SIMS and AES, showed the value of using complementary techniques.

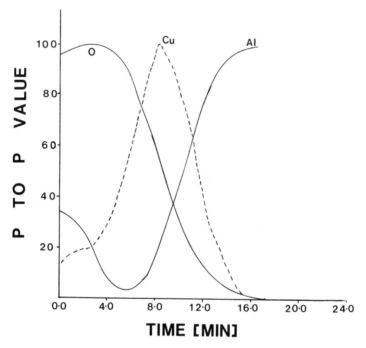

FIG. 7. Oxygen, aluminium and copper elemental profiles with depth for 2024 Al alloy treated with FPL etch.

Sometimes various processing steps on aluminium alloys result in formation of 'smut', a loosely adhering, powdery material on a surface which makes it unsuitable for bonding. One such smut was formed on 2024 aluminium alloy with a standard chromate stripping solution. Figure 8 shows ISS and SIMS data from the smutted surface compared with an area which had been above the solution and had not become smutted. The ISS spectrum of the smut shows it to be largely copper; similarly, SIMS shows a high concentration of copper. Because of the low yield of the positive ions for copper and the high yield for chromium, the SIMS spectrum from the smut appears to show a high concentration

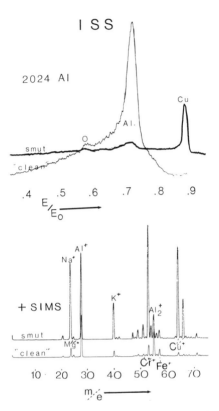

FIG. 8. ISS and SIMS data from a smutted surface of 2024 Al alloy compared with non-smutted surface.

of chromium. Such spectra must be carefully interpreted in order to estimate concentrations at the surface. Neither ISS nor positive SIMS is sensitive to carbon, especially in finely divided form, on the surface. Auger electron spectroscopy, however, has extreme sensitivity for carbon. AES shows the same smutted surface to contain, in addition to copper and the other elements shown in the SIMS spectrum, a high concentration of carbon.

Even intermediate chemical steps may influence the final surface composition of an alloy. An example of this is seen in Fig. 9, where the aluminium alloy 7050 (nominal zinc 6·2%, copper 2·3%, magnesium 2·2% and zirconium 0·12%) was treated using an HF/HNO_3 solution which left fluorine on the surface. Surprisingly the fluorine remains on the

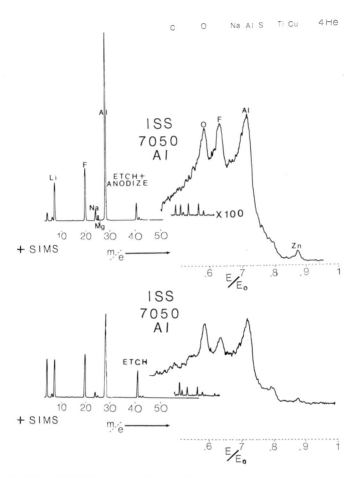

FIG. 9. ISS and SIMS data for 7050 Al alloy treated by HF/HNO₃ (bottom) and
then anodised (top).

surface during anodic oxide growth to produce a different stoichiometry
from that expected in pure aluminium oxide. FPL etch applied to this
same alloy results in the surface becoming zirconium rich even though
the bulk concentration of zirconium is only 0·12%.

There is an ever increasing demand for materials which can be used at
high temperatures with good corrosion resistance. Such requirements are
satisfied in many cases by stainless steels or titanium alloys.
Conventional methods of joining these materials such as welding, braz-

ing and soldering may be used, but reliable bonding methods are also being developed. Stainless steel following final processing is normally covered with a surface layer containing processing aids and oxidation products. The surface must undergo an etch or pickle before it may be used for the final processing. It has been found that generally this surface layer in its original state has a deleterious effect on adhesive bond strength.[15] Unfortunately, after acid pickling the surface can be even more contaminated then it was originally. This contamination, as in aluminium, is called smut. The usual method for desmutting is to wipe the work piece after rinsing, while it is still wet, or to brush mechanically with a stiff wire bristle brush. However, it has been found that only a portion of the smut is removed by these methods and chemical etches are required to remove all of the material.

An example of smut on stainless steel is shown in Fig. 10, where a surface which had been treated in hot sulphuric acid is shown. The elements which are observed are silicon and oxygen with some carbon present on the surface and only a slight amount of the matrix stainless steel. The SIMS spectra were taken simultaneously with the ISS data and show primarily silicon and cluster peaks of silicon and oxygen together with a small amount of the matrix material. Even when the surface of the 304 stainless steel was wiped after rinsing and appeared visibly clean, some smut remained behind as detected by both positive SIMS and ISS. The smut was completely removed by dipping the work in hot caustic soda solution or in chromic acid solution conventionally used in the

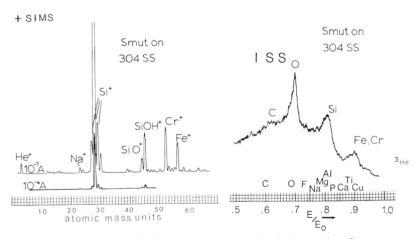

FIG. 10. ISS and SIMS data for a smutted stainless steel surface.

industry. Again, as was shown with the case of the aluminium, there is great advantage to using several techniques for analysis. The analysis of the smutted surface and the desmutted surface by a variety of techniques is shown in Table 5. Each of the techniques has its strong points. In AES, for instance, both sulphur and chlorine are readily detected in the spectrum, whereas in the ISS spectrum neither of these elements was detected, probably because they occur in an area of high scattered ion background. The XPS data show definitely that silicon is in an oxidised state which would be expected from the formation in the highly oxidative medium of the acid etch.

TABLE 5

SPECIES FOUND ON 304 STAINLESS STEEL BY SURFACE ANALYSIS TECHNIQUES

Technique	Smutted in sulphuric acid	Desmutted in sulphuric–chromic acids
AES	Si, O, S, Cl C, Cu, Fe, Cr, Ni	O, C, S, Fe, Cr, Ni
ISS	Si, O, C, Cu(?) Fe, Cr	O, Na, Fe, Cr
+SIMS	Si^+, Na^+, SiO^+, S, OH^+, Cu^+, CH_n Fe^+, Cr^+, Ni^+, Cu^+	Na^+, K^+, Si^+, S^+, OH^+ CH_n^+, Fe^+, Cr^+, Ni^+
−SIMS	$C_nH_n^-$, O^-, OH^- Si^-, SiO^-, SiO_2^- SiO_3^-, SiF^-, $SiOF^-$ SiO_2F, FeO_2^-	$C_nH_n^-$, O^-, OH^-, Cl^- SiO_n^- (greatly reduced) CrO^-, CrO_2^-, CrO_3^- FeO_2^-
XPS	C,[a] O, S, Si,[b] N, Cu, Fe, Cr, Ni	C, O, Fe, Cr, Ni

[a] More than one form
[b] Oxide form

Similar contamination or the formation of smut may occur on the surface even when the specimen does not contain elements which cause such a condition; for instance, pure titanium was found to smut in a solution of nitric acid and hydrofluoric acid. The ion scattering spectrum shown at the bottom of Fig. 11 shows that in addition to titanium and oxygen there is also copper present. The history of the acid solution was traced and was found to have been used earlier to etch 2024 aluminium

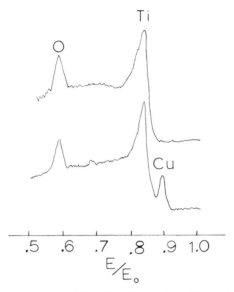

FIG. 11. ISS data for a smutted Ti surface (bottom) and for a clean surface (top).

alloy which contains copper. When a fresh solution of the same concentrations of these two acids was made up the resulting etched surface was clean and contained no copper, as shown in the upper spectrum of Fig. 11. Such unambiguous results may not be obtained using only SIMS, as shown in Fig. 12. Here the copper isotopes (63 and 65) occur at the same nominal mass as the TiO^+ series of ions. Again the unmistak-

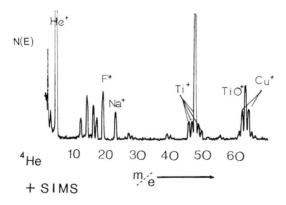

FIG. 12. SIMS data for smutted Ti surface.

able advantage of using more than one surface characterisation is evident.

Surface characterisation methods are useful in determining the degree of cleanliness of adherend surfaces. The combination of short wave length ultraviolet light plus ozone, such as that produced by a quartz ultraviolet lamp, is one method of cleaning and storing adherend surfaces.[16,17] Typical ISS–SIMS data are shown in Fig. 13 for Ti-6Al-4V as received (Ti-6Al-4V). This specimen was not degreased or treated in any way. The ISS spectrum is very weak showing many elements and in particular carbon. Large quantities of sodium and potassium and fluorine, apparently from the processing of the alloy, are observed in addition to the alloy constituents. Note the cluster spectra; the calcium fluoride ions CaF^+ may indicate that the fluorine is tied up on the

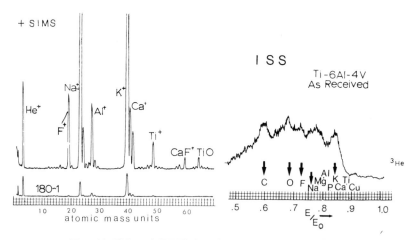

FIG. 13. ISS and SIMS data for as received Ti-6Al-4V.

surface by calcium. Titanium is hardly visible in either the ISS or SIMS data. Figure 14 shows data from the same specimen cleaned under ultraviolet light. The ISS spectrum now shows primarily titanium and oxygen; perhaps a small amount of carbon still exists but upon cleaning for a longer period of time all trace of carbon disappears. The ISS spectrum is now much stronger, taken on a full scale of 25 000 counts per second as compared to 5 000 counts per second in the earlier, as received, specimen. Note also the change in the SIMS data where before titanium was barely visible and now is very strong. The TiO^+ cluster spectra are

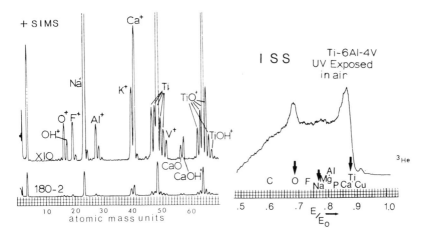

FIG. 14. ISS and SIMS data for the specimen used in Fig. 13, but cleaned with UV/O$_3$.

very strong, indicating a great deal of oxygen activity on the surface. Note also the appearance of O$^+$ and OH$^+$ relative to the fluorine that had been present initially. The CaF$^+$ has disappeared and now cluster ions CaO$^+$ and CaOH$^+$ appear in the spectrum. Notice also the dramatic decrease of the potassium positive ion yield on this surface, compared with calcium for instance. No explanation is offered for this apparent decrease except that perhaps a change in chemical state produced a large relative yield change.

4.2. Analysis of failure mode

The strength of an adhesive joint is measured by means of numerous physical tasks which place the joint in shear or tension or a combination of the two. These tests, in which an increasing load is placed on the joint until failure occurs, give some idea of the initial bondability of an adhesive–adherend combination. Similar tests in which the bond is under load but at high temperatures and humidity are accelerated tests of bond durability. In the past, following joint failure, visual or sometimes microscopic examination of the failure surfaces was made to determine the mode of failure. A major consideration in identifying the mechanics of adhesive joint failure is the locus of fracture, whether the joint failed by cohesive fracture in the adhesive, or adhesive failure interfacially between the adhesive–substrate interface, or a complex mixture of possible failure modes. A long time theory of Bikerman[18] says that true interfacial failure occurs so seldom that this failure mode need not be treated in any theory of adhesive joints. He says that apparent failures in adhesion are quite common, but they take place in a weak boundary layer so near the interface that the adhesive remaining on the adherend is not visible. Such failure in a weak boundary layer has been analysed in this laboratory using ISS–SIMS.[19,20] In this work it was shown that, when scanning electron microscopy (SEM) and spectrochemical tools are used to determine the morphology and the chemical species on the surface, there still may be difficulties in interpreting the location of failure. Some failures are very clearcut while in others, particularly mixed mode failures, interpretation may not be as easy. Figure 15 shows the schematic diagram of a typical complex adhesive bonded system in which several interfacial regions exist. Each of the materials, coming together to form the interface, has its own individual chemical signature. The location of the failure may be precisely determined from these individual signatures. The substrate, for instance, usually contains alloying elements which vary in content between the surface and the bulk. In addition to alloying elements, surface treatments leave behind elements characteristic of each treatment, for example, the popular etch used for aluminium alloys consisting of sulphuric acid and sodium dichromate leaves a detectable amount of chromium on the alloy surface. Primers often contain anions or cations which can be followed by spectrochemical methods. Additives such as strontium chromate are usually placed in

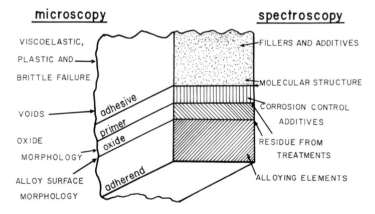

FIG. 15. Schematic diagram of adhesive bonded system.

the primer to provide corrosion protection in the coating. An example of SIMS data from a peel test specimen which failed along the adhesive–primer interface is shown in Fig. 16. The SIMS data as well as ISS spectra show unmistakably that strontium is present on the adherend side of the failure. In order to make a full interpretation of the locus of failure, it is necessary to examine both sides of the failure. The failure surface which included the adhesive also shows strontium, suggesting that the failure was a mixed mode failure.

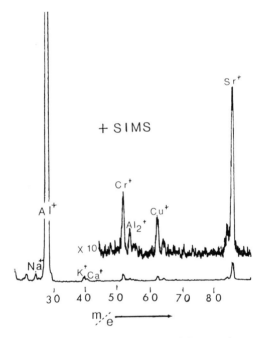

FIG. 16. SIMS data for peel test failure surface.

Failure surfaces from thin adherend wedge test specimens have proved to be an interesting illustration of the use of ISS–SIMS surface character-isation. The wedge test method provides information about adherend surface preparation. This test configuration is sensitive to different surface preparation treatments and can discriminate between bonding processes that give good and poor service performance. The wedge specimen consists of two thin adherends with a wedge driven into the bond line as shown in Fig. 17. The position of the crack leading edge is

W. L. BAUN

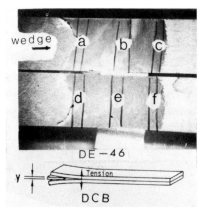

FIG. 17. Schematic diagram of the wedge test specimen and a typical failed test specimen.

determined microscopically and then the wedge test specimen is subjected to various external stimuli such as changes in temperature and relative humidity. The propagation of the crack tip is followed with time. Sometimes when the wedge is driven into the bond line, separation of the specimen occurs over a portion of the bond line as shown: separation first occurs as cohesive failure in the adhesive at the left, then with increase in temperature and humidity there is a change in the failure mode during testing, with the failure going to the interface. The areas shown in the figure as A, B, C (clean adherend) and D, E, F (clean adhesive) are matching regions from a wedge test specimen which failed. The adherend area which failed during the actual testing at high temperature and humidity shows no indication of adhesive either visually or in the SEM although there are slight reflectivity differences along the surface of the aluminium. ISS and SIMS data for the adherend surface of this specimen are shown in Fig. 18. These spectra indicate substantially adhesive (interfacial) failure at the oxide–primer interface at A and B. The chromium content in some areas may indicate some cohesive failure in the oxide since the original adherend was prepared in a standard FPL (Forest Products Laboratories) etch which contains chromate ion. The matching 'clean' adhesive side gave the ISS–SIMS data shown in Fig. 19. Spectra at D and E appear to be contaminated primer or a mixed primer adhesive zone while F appears to come from a contaminated film of oxide which has adhered to the adhesive. The appearance of aluminium even at position D seems to indicate some cohesive failure in the oxide or

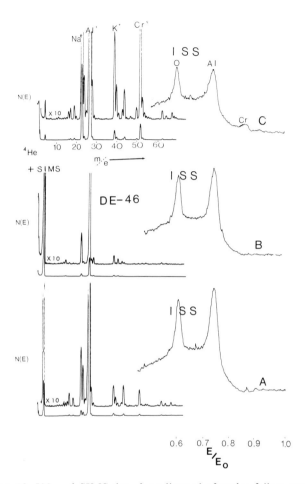

FIG. 18. ISS and SIMS data for adherend of wedge failure test.

in the oxide–metal interface all the way across the clean region of the specimen. ISS–SIMS data from other specimens indicate convincingly that failure sometimes occurs in the oxide or at the oxide–metal interface. Analysis of several such fractures after exposures to high humidity and elevated temperatures suggest early crack propagation at the primer–oxide interface with large amounts of impurity ions such as sodium and potassium present in this region. With continuing time, the crack appears to progress either cohesively in the oxide or adhesively at the oxide–metal interface. The important fact is that all surface treatments do not

W. L. BAUN

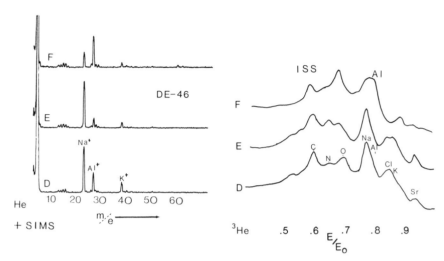

FIG. 19. ISS and SIMS data for matching adhesive side from wedge test in Fig. 18.

behave in this way and therefore the wedge test is a useful indicator of the durability of a particular surface treatment for aluminium. Likewise, an analysis of the locus of failure using such sensitive surface techniques as ISS and SIMS makes the wedge test even more useful to evaluate surface treatments.

Ion spectroscopy techniques were also very useful when applied to failure surfaces in a fundamental study of adhesion of gold on Ti-6Al-4V. Ti-6Al-4V alloys were etched for varying periods of time in stirred solutions of hydrofluoric acid and ammonium phosphate. Following etching, one half of the specimens was covered with vacuum evaporated gold while the other half was bonded with commercial adhesive. Gold adhesion to the alloys was evaluated by a peel test and a lap shear test using the same commercial adhesive. The uncoated specimens were evaluated by lap shear test. Adhesion of Ti-6Al-4V to gold and to the commercial adhesives was directly proportional to the length of time the specimen was etched. In mixed failures such as those shown in Fig. 20, ISS gave a quantitative measure of how much gold remained on each surface. Figure 20 shows ISS spectra from the original Ti-6Al-4V surface (a) and from the surface after stripping of the gold (b). Note that in addition to the gold, there are large amounts of the alkali ions present on this surface. Surface chemistry changes as determined by ISS–SIMS suggest selective etching of the alpha phase in the alloy and subsequently

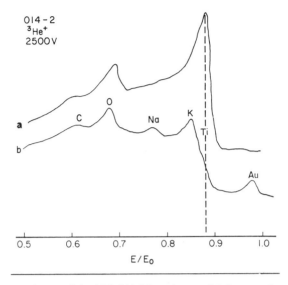

FIG. 20. ISS Data from original Ti-6Al-4V surface and failure surface following stripping of gold.

the large influence of the beta phase in bonding. Heating of the gold on Ti-6Al-4V resulted in improved adhesion, probably by diffusion mechanisms. Exposure to steam resulted in bond degradation in both gold/Ti-6Al-4V and in adhesive/Ti-6Al-4V systems. This perhaps suggests degradation due to the oxide interface. Adhesive bonding results for the etched specimens were compared with those expected based on attachment site theory.[21] Excellent agreement for both gold on Ti-6Al-4V. and the commercial adhesive on Ti-6Al-4V was observed. Degradation of the bond due to steam follows the same form in both systems.

5. CONCLUSIONS

Ion spectroscopy methods used with other techniques such as SEM provide powerful tools for elemental analysis of surface composition in adhesive bonding research. Such techniques show that surface composition can be significantly different from the bulk composition due to contamination, selective chemical etching and segregation. Such chemical inhomogeneities in the interphase region can have a significant effect on adhesive bonding. Ion spectroscopy techniques are also invaluable in providing an analysis of the mode of failure in adhesive joints. It is necessary to make such an analysis on both sides of the failure surface in order to understand fully the mechanics of such a failure. Visual inspection of a failure surface frequently results in the classification of 'adhesive' failure. Often mixed mode failures or failures in a new interface containing elements of both adhesive and adherend actually occur and can be deduced by the use of modern surface characterisation methods.

REFERENCES

1. Buck, T. M. in *Methods of surface analysis*, A. W. Czanderna (Ed.), Elsevier, Amsterdam 1975, p. 75.
2. Smith, D. P., *J. Appl. Phys.*, **38** (1967), 340.
3. Chen, Y. S., Miller, G. L., Robinson, D. A. H., Wheatley, G. H. and Buck, T. M., *Surf. Sci.*, **62** (1977), 133.
4. Winters, H. F. in *ACS Adv. Chem. Ser*, **158**, M. Kaminsky (Ed.), (1976).
5. Benninghoven, A., *Surf. Sci.*, **53** (1975), 596.
6. Werner, H. W. in *Electron and ion spectroscopy of solids*, L. Fiermans, J. Vennik, and W. DeKeyser (Eds.), Plenum Publishing, New York 1978, p. 324.
7. Werner, H. W., *Surf. Interface Anal.*, **2** (1980), 56.
8. Chu, W. K., Mayer, J. W. and Nicolet, M. A., *Backscattering spectrometry*, Academic Press, New York 1978.
9. White, C. W., Simms, E. B. and Tolk, N. H., *Science*, **177**, (1972), 481.
10. Baun, W. L. in *Adhesion measurement of thin films, thick films, and bulk coatings*, K. L. Mittal (Ed.) ASTM, Philadelphia, Pa. 1978, p. 41.
11. Rusch, T. W. and Erickson, J., *Vac. Sci. Technol.*, **13** (1976), 374.
12. Baun, W. L., *Appl. Surf. Sci.*, **1** (1977), 81.
13. Solomon, J. S. and Baun, W. L. in *Surface contamination, genesis, detection and control*, Vol. 2, K. L. Mittal (Ed.), Plenum Publishing, New York 1979, p. 609.
14. McDevitt, N. T., Baun, W. L. and Solomon, J. S., *J. Electrochem. Soc.*, **123** (1976), 1058.
15. Allen, K. W. and Alsalim, A. S., *J. Adhes.*, **8** (1977), 183.
16. Vig, J. R. in *Surface contamination, genesis, detection and control*, Vol. 1, K. L. Mittal, (Ed.), Plenum Press, New York 1979, p. 235.
17. Baun, W. L., *Appl. Surf. Sci.*, **6** (1980), 39.
18. Bikerman, J. in *Recent advances in adhesion*, L. -H. Lee (Ed.), Gordon and Breach, New York 1973, p. 351.
19. Baun, W. L., *J. Adhes.*, **7** (1976), 261.
20. Baun, W. L. in *Characterization of metal and polymer surfaces*, Vol. 1, L. -H. Lee (Ed.), Academic Press, New York, 1977, p. 375.
21. Lewis, A. F. and Natarajan, T. T. in *Polymer science and technology*, L. -H. Lee (Ed.), Plenum Press, New York 1975, p. 563.

Chapter 4

CHEMICAL ANALYSIS OF POLYMER SURFACES

D. Briggs

ICI Petrochemicals and Plastics Division, Welwyn Garden City, UK

1. INTRODUCTION

The chemical analysis of polymer surfaces shows some of the features of metal surface analysis described in earlier chapters. However, the very different nature of organic polymeric materials and metals inevitably leads to many differences in both analytical philosophy (what kind of information is thought to be important?) and approach (which techniques can supply this information?)

Perhaps the major distinguishing feature of polymer surfaces is their variable morphology. Semi-crystalline polymers are obviously heterogeneous, possessing both crystalline and amorphous phases. Copolymers (particularly block copolymers) may be composed of discrete domains with different chemical composition. Individual polymer chains may exist in states of differing orientation. Although these may be regarded as aspects of physical structure they nevertheless *could* play an important role in adhesion to polymer surfaces, either directly or indirectly through their influence on pretreatment effects. For this reason, morphological aspects of structure must be encompassed within 'chemical' analysis of polymer surfaces.

Of the techniques described in Chapters 2 and 3, only XPS (ESCA) has so far made any impact on polymer surface analysis. Organic materials are very easily damaged by electron beams and in conventional Auger electron spectroscopy (AES) the beam current density required rapidly leads to polymer pyrolysis. In any case, charging effects with these highly insulating materials would cause severe problems even at lower current densities. Thus, high spatial resolution chemical analysis of polymer surfaces is made difficult, if not impossible. To date, there has been very little application of ion spectroscopy (SIMS or ISS) to polymer analysis. Organic materials are more sensitive to ion beam damage than are inorganics; however, initial studies suggest that meaningful spectra (i.e. of undamaged surfaces) can be obtained from subcritical exposures. ISS in particular has considerable potential. Depth profiling by ion sputtering is widely used in metal surface analysis although chemical state analysis as a function of depth is complicated by ion-induced damage (reduction, rearrangement, etc). With polymers, the situation is even more complex because the sputtering process is not expected to be smooth and because compositional changes occur with much greater facility. This situation is particularly disadvantageous when interface analysis is required.

A major additional source of chemical information from polymers comes from infrared and Raman spectroscopy methods. These vibrational data can be obtained from the surface region, although the information depth is much greater than with the above techniques. Reflection IR techniques are very widely used and surface sensitivity has been significantly improved by the advent of Fourier transform (interferometry) instruments. Raman spectroscopy in the form of the laser–Raman microprobe promises a reasonable degree of spatial resolution in polymer surface analysis. These techniques have the advantage that they can also provide information on polymer morphology.

The rest of this chapter will concentrate on a discussion of XPS and reflection IR, being the techniques which have contributed most to the chemical analysis of polymer surfaces. Ion spectroscopy and Raman microprobe analysis will be discussed more in the context of how they may contribute in the future.

2. X-RAY PHOTOELECTRON SPECTROSCOPY

The instrumentation for, and physical basis of, XPS (ESCA) has been described in Chapter 2. To avoid unnecessary repetition only those features which have a direct bearing on polymer surface analysis will be discussed in this section.

2.1. Sample handling
No special facilities are required for polymer work in general. Contrary to the expectation of many workers who are used to studying metallic

systems, polymer samples do not present serious outgassing problems and can easily be introduced into high vacuum systems attached to the sample probe by means of double-sided adhesive tape (itself polymeric!). In the author's experience, polymer surfaces, even those of systems containing additives, are reasonably stable (in the sense that spectra are not time dependent to a significant degree). Prolonged exposure to X-ray beams of typical flux density will often produce visual evidence of damage (discolouration), but few polymers undergo rapid change. Of the common polymers, those containing chlorine (especially poly(vinyl chloride), PVC) seem to be most sensitive.

Contamination of polymer surfaces does not constitute a major problem provided some care is taken, In general, the sticking coefficients of residual gas molecules on polymer surfaces are rather low (cf. on metals); nevertheless, the build-up of hydrocarbon-like material is frequently observed within the timescale of the experiment. This affects quantification, particularly if low KE peaks (involving electrons with low escape depth) are to be measured, but perhaps more importantly it affects directly the C1s profile. This contamination almost certainly comes from the X-ray gun casing or window in close proximity to the sample. Working with polymers routinely leads to this problem because the spectrometer working pressure is usually significantly greater than base pressure over extended periods due to slow outgassing and loss of volatile low molecular weight constituents. The best preventative measure is regular system bake-out. Since the X-ray gun heats up, samples being studied for long periods can get quite warm (particularly thin film samples) and thermal effects may come into play (surface degradation, migration and/or loss of additives, etc). Cooling the sample holder may overcome these effects.

2.2. Instrumentation

A fundamental problem in polymer surface analysis is the occurrence of charging effects due to the insulating nature of the materials. Usually the charging effect is less that 5 V, and as might be anticipated, the magnitude is a strong function of surface composition.[1] When monochromation of the X-ray beam is employed the situation is much worse (largely because the bremmstrahlung radiation is removed, thereby greatly reducing secondary electron emission). In this case, in is mandatory that some form of charge neutralisation is used—usually a 'flood gun' to provide a low energy electron flux over the sample surface.

In general, the preferred excitation source is MgKα since this leads to the smallest peakwidths. A dual-anode source (giving either MgKα or AlKα) is useful in cases where overlap occurs between a core level of interest and an X-ray excited Auger peak. Switching anodes moves all the photoelectron peaks in the spectrum by 233 eV without affecting the Auger peaks (detecting low levels of sodium in chloropolymer samples using MgKα, for example, where Na1s and ClLMM are closely spaced). The usefulness of X-ray monochromation (perforce with AlKα) in studying polymers is debatable. Compared with using MgKα under the highest resolution conditions the improvement in resolution is not very significant and very careful charge compensation is required before this can be achieved. Data collection times are longer (reduced X-ray flux) but X-ray satellites which can be a spectral nuisance[2] are removed. On the other hand, data processing of normal (e.g. MgKα excited) spectra can perform this function very quickly, assuming that the required computer is on hand.

One particular instrumental facility which stands out in polymer surface analysis is the ability to change the angle between the sample surface and the analyser entrance slit. If this angle is θ (the 'take-off' angle) then the information depth decreases as sin θ. Thus, at low take-off angles the 'surface sensitivity' increases (see section 2.4).

2.3. Information from XPS

2.3.1. *The information depth*
Although early studies seemed to indicate longer inelastic mean free paths (IMFP) for electrons (of a given KE) in organic systems than in metals and inorganics, a different view on polymer systems is now emerging. In general, it seems that a 'universal curve' for IMFP[3] in all materials could reasonably be drawn to include polymers.[4] Thus, the sampling depth ($\simeq 3\lambda$ where λ is the appropriate IMFP) for polymers varies approximately between 15 and 100 Å for electrons of KE from 100 to 1500 eV.

2.3.2. *Core level spectra*
Core levels of elements commonly encountered in polymers are readily observed, but in bulk terms the technique has a detection limit of only $\simeq 0.5$ at%. Signal intensities can be quantified using relative sensitivity factors appropriate to the instrument used (if known), [5-7] determined *in situ* (using homogeneous standards), or arrived at using analyser transmission function data and factors from other instruments.[8] Some repre-

sentative data are shown in Table 1. Good examples of the attainable quantification precision are to be found in reference 9.

In many instances, the ability to detect and quantify certain elements on, or near, the surface of a polymer is sufficient to satisfy a point at issue. Frequently, however, structural information is required (how the atom in question is bonded) and core level chemical shifts might be useful. It should be noted immediately that carbon bound to itself and/or to hydrogen has the same 1s binding energy (BE) no matter what the state of hybridisation. It requires the attachment of more electronegative atoms or groups to induce measurable chemical shifts. Some examples are given in Table 2.

TABLE 1

RELATIVE SENSITIVITY FACTORS (C1s = 1·00) FOR MgKα EXCITATION

	Kratos ES 200		Varian IEE-15[c]	
	FRR[a]	FAT[b]	Peak area	Peak height
C1s	1·00	1·00	1·00	1·00
N1s	1·19	1·39	1·71	1·56
O1s	1·67	2·40	2·54	1·93
F1s	1·92	3·74	4·17	3·70
Si2p	1·20	0·97	0·71	0·92
S2p	1·89	1·63	1·38	1·70
Cl2p	2·44	2·19	1·92	2·03

[a] Peak area. Fixed retardation ratio mode where E_s/E_0 [E(ejection KE)/E(transmission)] = 0·23. From ref. 7.
[b] Peak area. Fixed analyser transmission mode $E_0 = 65$ V. Calculated from FRR data using $I_{FAT}/I_{FRR} = E_s^{-1·25}$ (ref. 35).
[c] FAT mode $E_0 = 100$ V. From ref. 5.

It will be noticed that the dynamic range of these shifts is rather small, that is the maximum shift divided by the peakwidth (at half-height), is ~ 10. Moreover, small secondary effects induced by neighbouring groups produce a small spread in BEs for any single functional type. What this amounts to in practice is that, for analysis of a mixture of functional groups, even involving only carbon and one other element, XPS data lack precision. For instance, for a surface in which only oxygen and carbon are known to be present, a C1s peak with BE ≃ 286·5 eV (relative

TABLE 2

C1s CHEMICAL SHIFTS RELATIVE TO CH₂—$\underline{C}$H₂—CH₂

Structure	ΔC1s (eV)			Ref.
$\underline{C}$F₃—$\underline{C}$H₂—CH₂	7·0		1·2	37
CF₂—$\underline{C}$F₂—CF₂		7·2		38
CH₂—$\underline{C}$F₂—CH₂	5·8		0·7	38
CH₂—$\underline{C}$HF—$\underline{C}$H₂	3·0		0·4	38
CF₂—$\underline{C}$HF—CF₂		4·3		38
CH₂—$\underline{C}$HCl—$\underline{C}$H₂	1·3		0·2	37
CH₂—$\underline{C}$H(OH)—$\underline{C}$H₂	1·6		0·2	37
CH₂—$\underline{C}$(=O)—$\underline{C}$H₂	3·1		0·5	37
$\underline{C}$H₂—$\underline{C}$OOH	0·7		4·7	37
$\underline{C}$H₂—$\underline{C}$OO$\underline{C}$H₂	0·6	4·4	1·6	37
$\underline{C}$H₂—O—CH₂		1·8		37
CH₂—O—$\underline{C}$(=O)—O—CH₂		5·8		9
CH₂—O—$\underline{C}$H₂—O—CH₂		3·0		9
$\underline{C}$H₂—$\underline{C}$H₂—C≡N	2·1		2·1	37
CH₂—$\underline{C}$H₂—NH₂		0·9		37

to C1s [hydrocarbon] $= 285\cdot0$ eV) could arise from

$$>\text{C}-\text{OH}, \quad >\text{C}-\text{O}-\text{C}<, \quad >\text{C}-\text{OOH}, \quad -\overset{\displaystyle \overset{\text{O}}{\|}}{\text{C}}-\text{O}-\underline{\text{C}}, \text{ etc}$$

O1s in organic systems has a much smaller dynamic range that C1s ($\cong 2$) so combining O1s BE data does not help much. The N1s dynamic range is similar to that of C1s and identification of nitrogen functions is slightly easier.

A multifunctional surface will give rise to a broad C1s spectrum containing overlapping peaks. Frequent recourse is made to deconvolution (curve resolution) methods in order to identify and quantify the components. Since a unique solution can rarely be guaranteed, this procedure is fraught with uncertainty and the outcome should be treated with due caution. A relatively simple example is shown in Fig. 1.

In order to overcome this lack of binding energy specificity, attention has recently turned towards derivatisation methods.[10-12] Here, the aim is to introduce unique elements (which can easily be identified and quantified) via a reagent which 'derivatises' a specific functional group, e.g.

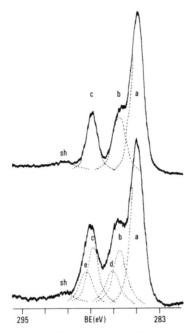

FIG. 1. Deconvoluted C1s profiles from poly(ethylene terephthalate) obtained using a Du Pont 310 Curve Resolver. Upper trace: untreated PET, (a) —$\underline{C}_6H_4$—;

(b) —$\underline{C}H_2O$—; (c) —$\overset{\overset{O}{\|}}{\underline{C}}$—$OCH_2$; (sh) shake-up satellite ($\pi \to \pi^*$). Lower trace: electrical discharge treated PET with additional peaks assigned to (d) $\underline{C}$—OH (phenolic) and (e) —$\underline{C}OOH$. (see also ref. 36.)

$$-COOH + NaOH \longrightarrow -COO^- Na^+ \quad \text{(Na label)}$$

In many cases, it can also be arranged that sensitivity is enhanced by using labels with high sensitivity factors (e.g. for $>C=0$, the use of pentafluorophenylhydrazine gives an amplification of $\times 10$). More details of these procedures are given in Chapter 9.

In cases where the polymer contains unsaturation (e.g. polyacetylene) and particularly when this involves aromaticity (e.g. polystyrene), shake-up satellites can be detected on the low KE (high BE) side of the C1s peak. These arise from simultaneous core electron emission and an 'optical' transition of an electron from a filled molecular orbital to a higher unfilled molecular orbital (e.g. $\pi \rightarrow \pi^*$).[13] The separation of these peaks and their relative intensity can cause problems through overlaps with high BE C1s peaks (introduced, for example, by oxidation). Shake-up satellites are, however, a direct monitor of aromaticity (see Fig. 1).

2.3.3. Valence band spectra
Photoemission from the molecular orbitals (i.e. of electrons involved in chemical bonds) also occurs in XPS, but with much lower probability than from core levels under normal conditions (viz. MgKα or AlKα excitation). The resulting valence band (VB) spectrum obviously contains information on primary polymer structure, but sophisticated calculations are required to elucidate the correlation.[14] However, VB spectra have fingerprint value. Figure 2 shows this for three polyolefins which give identical core level (C1s) spectra.

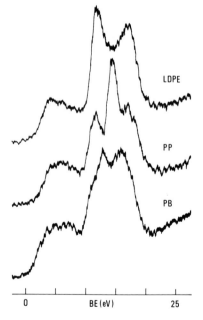

FIG. 2. XPS valence band spectra of low density polyethylene (LDPE), polypropylene (PP) and polybut-1-ene (PB).

2.4. Depth resolution

Non-destructive techniques for depth resolution can only give information on profiles within the maximum sampling depth, which is dependent on the exciting radiation and core levels available. Measurement of relative signal intensities as a function of electron take-off angle is often rewarding. The principle is shown in Fig. 3. Relative

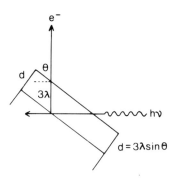

FIG. 3. Variation in vertical sampling depth (d) of XPS as a function of the electron 'take-off' angle (θ). In the particular case illustrated, the angle between the X-ray beam and the analyser entrance slit is fixed at 90°.

core level intensities for a sample homogeneous within a depth of $3\lambda_{max}$ (calculated for the peak with highest KE), are invariant to change in θ. As θ decreases, an inhomogeneous distribution shows up as an increase in relative intensity of peaks arising from species closest to the surface. An example is shown in Fig. 4. It should be emphasised that this technique is only applicable to smooth surfaces. Another technique is to monitor the relative intensities of widely spaced (in KE) core levels of the same element. For an homogeneous sample (as defined above) this ratio has a certain value. Ratios greater than this arise when the element is distributed in a surface layer thinner than $3\lambda_{max}$. This technique has been successfully used with F1s/F2s[15] and O1s/O2s combinations[16] (see also Chapter 9). Destructive depth profiling, by argon ion etching, has been little used. It is certainly not the routinely useful tool for polymer studies that it is for metal studies. This is because (a) the sputtering process is much more complex and (b) the polymer usually undergoes rapid structural change (e.g. cross-linking, elimination reactions via radical processes) even before skeletal fragments are lost.[17] Analysis of compositional changes below the sampling depth therefore requires a different approach, but this problem has not yet been tackled.

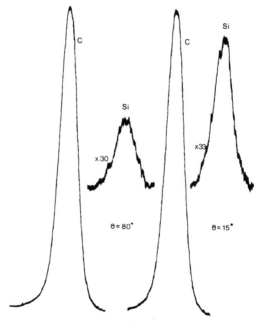

FIG. 4. C1s and Si2p signals from chromic acid etched LDPE contaminated with a silicon containing material (probably silicic acid). Note the marked increase in Si/C ratio as the surface sensitivity is enhanced by changing to a low take-off angle.

3. INFRARED SPECTROSCOPY

The examination of polymer surfaces by infrared spectroscopy relies, at present, almost exclusively on some form of reflection spectroscopy. Transmission techniques have been adapted in some cases; the stacking of very thin (solution cast) films of high surface:volume ratio has been described by Hayes;[18] surface layer removal by abrasion with KBr powder followed by pressing a disc of the abrasive plus material removed has been suggested by Johnson[19] and appraised by Willis and Zichy.[20]

3.1. Reflection IR techniques

3.1.1. *Principles*
For a thorough treatment of this subject the book by Harrick[21] is recommended. The basic technique requires that the infrared radiation is

internally reflected at the interface between a prism and the sample under investigation (as in Fig. 5a). During 'reflection', some penetration into the sample takes place so that, at energies where selective absorption occurs, the beam is attenuated. This (original) single reflection variant is consequently known as attenuated total internal reflection (ATR) infrared spectroscopy. Harrick[21] developed the following relationship for the effective penetration depth of the radiation (i.e. the sampling depth of the technique):

$$d_{\mathrm{p}} = \frac{\lambda_0}{2\pi n_1 [\sin^2 \theta - (n_2/n_1)^2]^{1/2}}$$

where d_{p} is defined at the distance below the surface at which the amplitude of the electric field is $1/e$ of its initial value, θ is the angle of incidence between the IR beam and the surface normal, n_1 and n_2 are the refractive indices of the reflecting element and sample respectively, and λ_0 is the wavelength of the radiation. Although n_1 varies only slightly with λ_0, n_2 is a strongly varying parameter. Figure 6 illustrates[22] its variation in the vicinity of an absorption band and typical ATR band shapes for certain relative values of n_1 and n_2. Clearly, to avoid distorted peak shapes, n_2 needs to be considerably greater than n_1(max.). Penetration depth decreases (or surface sensitivity increases) with decreasing λ_0, increasing θ and increasing n_1, but spectral intensity also decreases.

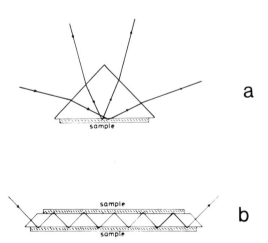

FIG. 5. Principles of attenuated reflection (ATR) and multiple internal reflection (MIR) infrared spectroscopy (a and b respectively).

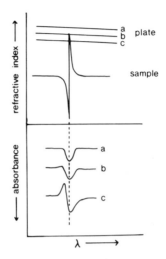

FIG. 6. Change in refractive index of a polymeric material in the vicinity of an IR absorption band and the shape of this band for three different relative values of sample and reflection plate refractive indices: (a) undistorted, (b) and (c) increasingly distorted. (Redrawn from ref. 22.)

Popular reflection elements are KRS-5 (TlBr–TlI mixed crystal, $n_1 = 2\cdot4$) and Ge. ($n_1 = 4\cdot0$). Table 3 gives some typical values of d_p using these materials (for polymers in general $n_2 \simeq 1\cdot5$).

Undistorted spectra from a minimised sampling depth using ATR have low intensity. Spectral intensity can be increased by increasing the number of reflections with the arrangement shown in Fig. 5(b). This is multiple internal reflection (MIR) spectroscopy. Common MIR attachments for IR instruments give nominally 9 or 25 reflections, with fixed or variable angle of

TABLE 3

MIR CONDITIONS AND DEPTHS OF PENETRATION

Reflection element	n_2/n_1^a	Angle of incidence	d_p/λ_0	d_p at 1723 cm^{-1} (μm)
Ge	0·378	45	0·067	0·39
KRS-5	0·631	60	0·122	0·65
KRS-5	0·631	45	0·208	1·21

[a]Ratio of refractive index of PE (1·515) to that of reflection element (Ge = 4·0, KRS-5 = 2·4). For most polymers $n_2 \cong 1\cdot5$.

incidence. In fact, the whole area of the plate–sample contact is sensitive, since the IR beam is not parallel and discrete reflections rapidly disappear as the rays travel down the plate. The plate area is typically $\sim 50 \times 20$ mm and the strength of the absorption spectrum can be altered by changing the dimensions of the sample.

3.1.2. Chemical information from reflection IR

From the viewpoint of qualitative analysis of organic systems in general, infrared spectroscopy reigns supreme. It is widely appreciated that the pattern of vibrational bands in the absorption spectrum (in terms of both frequency/wavelength and relative intensity) provides an identification 'fingerprint'. Libraries of characteristic IR spectra exist, including those of polymers.[23] In the context of this book, it is perhaps most important to discuss the identification of particular functional groups. Many compilations of characteristic frequencies are available; a particularly useful one is by White.[24] The specificity of these data is very dependent on which vibration is considered; some groups give very similar frequencies no matter what the molecular context, others can only be assigned to a certain range of frequency. It is also unusual to find precise relative intensities (usually vibrations are classified as strong, medium, weak, etc.) since the relevant physical quantity, the molecular extinction coefficient (ε) of the band, is very structure-dependent. It should be noted that ε values range over several orders of magnitude. For these reasons, quantification is very difficult, although it can be achieved.

A study which illustrates several of these points is that of Briggs and co-workers[16] on the oxidation of LDPE with CrO_3–H_2O. Figure 7 shows the changes which take place in the carbonyl stretching region. Characteristic frequencies to bear in mind are ~ 1710 cm^{-1} (carboxylic acid), ~ 1722 cm^{-1} (ketonic) and ~ 1735 cm^{-1} (carboxylic ester). Although peak overlap occurs it is clear that, as oxidation proceeds, the oxygen functions are dominated first by ketones and then by carboxylic acids with esters also being formed. For comparison with XPS data it was desired to compute the oxygen uptake as a function of time. Molecular extinction coefficients from model long chain hydrocarbons containing isolated ester, acid and ketone groups gave $\varepsilon = 350$, 500 and 160 respectively and, together with estimated relative intensities of the appropriate peaks within the carbonyl band, the percentage of oxygen present from each group could be calculated. By variation of d_p (in fact by using the conditions in Table 3) an oxygen concentration profile could be constructed. This is shown in Fig. 8. The points fall on a smooth curve

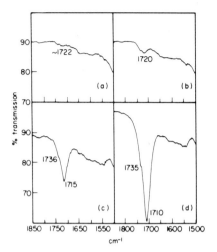

FIG. 7. MIR spectra of LDPE etched with CrO_3–H_2O at 25°C for (a) 0 s; (b) 20 s; (c) 1 h; (d) 5 h (ref. 16.)

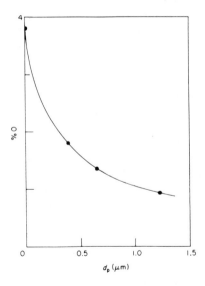

FIG. 8. Percentage of oxygen (atomic) in LDPE etched for 16 h at 25°C as a function of depth below the surface. Note that d_p defines the total depth over which the information is integrated by MIR. The point at $d_p \simeq 0$ is taken from XPS data (ref. 16.)

which extrapolates nicely to a 'surface' concentration given by XPS. A discussion of the form of concentration profiles and their effect on variable d_p studies is given in reference 25.

3.1.3. Morphological information from reflection IR

Orientation effects in polymer films can be studied by the use of polarised radiation. Willis and Zichy[20] have discussed the study of surface orientation of one-way drawn polypropylene film using polarised radiation and MIR. Their spectra are represented in Fig. 9. The direction of draw (machine direction, MD) and the plane of polarisation (direction of the electric vector, EV) are indicated by arrows. Vibrations along the molecular axis (in this case predominantly along the MD) give rise to *parallel* bands and these are more intense when the EV is parallel to this axis. When the EV is in the orthogonal direction (rotated by 90°) the *perpendicular* bands arising from vibrations perpendicular to the molecular axis are intensified. The behaviour of the bands in these spectra clearly indicate the high degree of MD orientation present in the sample studied.

Much work has been done on the assignment of IR bands from polymer chains in different configurations, e.g. alternative conformers, crystalline or amorphous phase. A study which illustrates both these aspects of morphology has also been reported by Willis and Zichy.[20] This concerns the chloroacetic acid etching of poly(ethylene terephthalate) (PET) film, a pretreatment used to improve adhesion. PET is a semicrystalline polymer with two molecular conformations, the extended (*trans*) and bent (*gauche*) forms. The latter exists only in amorphous regions whilst the former can exist in both crystalline and amorphous regions. The spectra of etched and untreated surfaces are shown in Fig. 10. Parallel and perpendicular bands are indicated (non-polarised light used) for both conformers, together with bands arising from chain folds in crystalline lamellae. Clearly, etching destroys crystalline PET; bands due to the *gauche* conformer increase dramatically in relative intensity and chain fold bands almost disappear.

3.2. Fourier transform IR

In conventional IR spectrometers, radiation from a polychromatic source interacts with the sample (suffering selective absorption) and is dispersed with a grating monochromator. The frequency range is scanned; that is to say only a fraction of the spectrum running time is used to collect intensity information at any particular frequency.

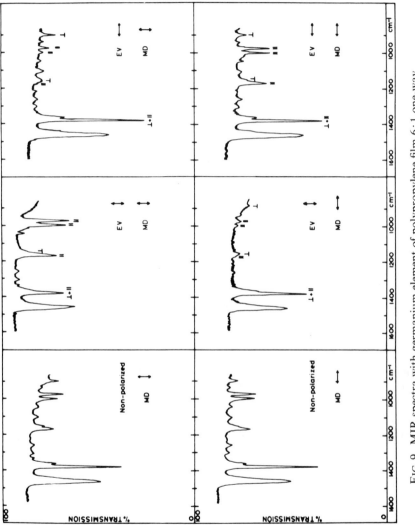

FIG. 9. MIR spectra with germanium element of polypropylene film 6:1 one-way drawn. Direction of draw (MD) and plane of electric vector (EV) indicated by arrows (ref. 20.)

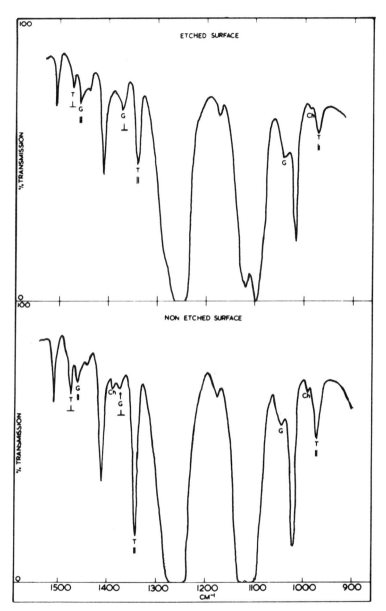

FIG. 10. MIR spectra with germanium element of PET film before and after chloroacetic acid etching. Owing to the strength of the spectrum; only one side of element was covered by the film. T = band due to *trans*-isomer, G = band due to *gauche*-isomer, Ch = band due to chain fold in crystalline lamellae (ref. 20.)

In Fourier transform IR the monochromator is replaced by an inter-
ferometer. A description of how the latter works is beyond the scope
of this chapter (see reference 26 for a succinct review); the essential
difference is that the interferogram produced *in any time* contains information on the intensity of each frequency in the spectrum. The extraction
of this information, i.e. to yield the IR spectrum, is carried out by the
mathematical operation 'Fourier transformation'. This is done by computer with digital information. All other things being equal, the
signal:noise (S/N) of spectra obtained by FT is M times greater than that
of spectra from a grating spectrometer, where M is the number of
resolution elements. However, for the same S/N FT spectra can be
obtained M times faster. Optical throughput is also much higher for FT
instruments. These advantages mean that spectral resolution can be
increased and that weak spectra can be more easily measured. Moreover,
the built-in computer can be used for data processing so that weak
signals can be further enhanced, e.g. by subtraction routines. Figure 11
illustrates some of these points. Although FT instruments are relatively
expensive, current developments are likely to improve their viability.

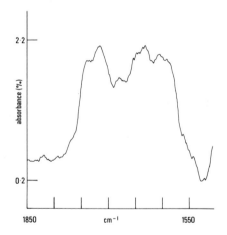

FIG. 11. FT-MIR spectrum obtained by subtracting spectrum of untreated
LDPE from that of LDPE treated in a nitrogen electrical discharge. Each
spectrum was an accumulation of 1000 scans ($\simeq 1$ per second) at $2\,cm^{-1}$
resolution. The spectrum is thought to contain contributions from carbonyl and
amine functions. Note the absorbance in this region—conventional MIR fails
to detect any significant change in the spectrum of LDPE after this treatment
(ref. 27).

4. ION SPECTROSCOPY

The principles of Ion Scattering Spectroscopy (ISS) and Secondary Ion Mass Spectroscopy (SIMS) have been outlined in Chapter 3. In the particular case of polymer surface studies, the highly insulating nature of the materials makes it essential that charge neutralisation is employed.[28]

To date, there have been very few reports of polymer surface studies by these techniques. However, these reports indicate a promising future for them. Gardella and Hercules[29] have obtained SIMS spectra from a series of alkyl methacrylate polymers. Although it might be anticipated that peaks at almost every mass number would be obtained, this work shows the mass fragmentation patterns to be characteristic and dependent on the nature of the alkyl side chain structure. The surface sensitivity of the technique was demonstrated by the ability to detect changes brought about by mild reactions (e.g. hydrolysis of poly(methyl methacrylate) by acetic acid). These same reactions caused negligible changes in the corresponding XPS spectra.

Gardella and Hercules[29] have also commented on the sensitivity of polytetrafluoroethylene (PTFE) to damage by ion beams, using ISS as the surface probe. The surface sensitivity of ISS is demonstrated by the fact that initial spectra of PTFE show only a peak due to F, i.e. the carbon skeleton is shielded. Ion damage was apparent (C detectable, O introduced) after 4.5×10^{16} ions cm^{-2} (Ne^+ at 2 kV).

Recent work in the author's laboratory suggests,[30] however, that the ion dose required to cause extensive surface damage to a polymer is much less than that inferred by Gardella and Hercules.[29] High quality spectra from polystyrene, easily interpreted in terms of conventional organic mass spectrometric data, have been obtained,[30] but it has also been shown[30] that significant changes in both absolute and, more importantly, relative peak intensities occur after $\sim 10^{13}$ ions cm^{-2} (Ar^+ at 4 kV). This corresponds to 27 minutes exposure to the rather low ion current density of lnA cm^{-2}. The spectra reported by Gardella and Hercules would correspond to surfaces accumulating 10^{13} ions cm^{-2} during the spectrum running time *alone*. Clearly great care is required for SIMS analysis of polymer surfaces, particularly when attempting to exploit the high spatial resolution potential of the technique.

The most extensive ISS study is that reported by Sparrow and Mishmash[31] concerning the surface treatment of 'Kaptan' polyimide. Surface contamination effects resulting from these treatments were most marked. In several cases, detailed depth profiles, to an estimated 30 Å below the surface, were made. These showed evidence of surface nitrogen depletion in many cases. An interesting conclusion from such a profile of a surface treated with tetramethylammonium hydroxide was that this molecule was being detected in the absorbed state with the methyl groups pointing away from the surface.

5. RAMAN SPECTROSCOPY

When monochromatic radiation strikes a material, a small amount is scattered. Most of this radiation has the same frequency as the original radiation (Tyndall or Rayleigh scattering) but a tiny proportion is scattered at discrete frequencies above and below this frequency (Raman scattering). Raman spectroscopy, as normally practised, is concerned with measuring the lower frequency (Stokes) bands; these contain vibrational information but are subject to different selection rules from normal infrared bands. Thus IR and Raman spectra are complementary.

The advent of laser sources has more than overcome the inherent problem of Raman spectroscopy, namely the weakness of the Raman scattered light, by providing intense monochromatic sources of radiation. Recently an experimental arrangement has been described for obtaining high quality Raman spectra of thin polymer films ($\sim 1\,\mu$m thick).[32] In the present context Raman spectroscopy offers an additional advantage—the laser light can easily be focused to a spot size limited only by the wavelength ($< 1\,\mu$m for laser frequencies in the visible region). The potential of this feature has recently been exploited in the Raman microscope or laser–Raman microprobe analyser.[33] The layout of this device is illustrated in Fig. 12.

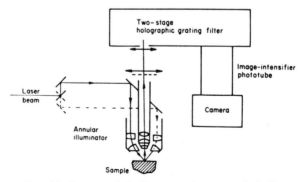

Fig. 12. Layout of the Raman microscope (ref. 33.)

Analyses (by indentification of characteristic Raman frequencies) can be obtained from a small selected area. However, by the use of multi-channel detection techniques (as opposed to single channel techniques employing a scanning monochromator), a larger area of sample can be illuminated (say 100 µm) and the image projected through an 'optical filter' which can be tuned to any single Raman frequency. In Fig. 12 this device is the holographic grating filter. The final recorded image is thus a map of those species in the illuminated area which give rise to the chosen Raman line. The relationship between optical microscopy and the Raman microprobe is therefore somewhat analogous to that between electron microscopy and scanning Auger microscopy.

This analogy might be pushed a bit further. One technique used for depth profiling in Auger spectroscopy is to polish a wedge shaped section. Spectra taken at points along this new surface correspond to points at varying depths below the original surface. By use of ultra microtome techniques, it should be possible to section polymers in this way so that the Raman microprobe can perform 'depth profiling'. Of course, the information depth of the technique is quite large; even with a beam focused on to the sample surface the depth of focus will be several microns. Nevertheless, since Raman spectroscopy, like IR, gives information on polymer morphology,[34] this could be a valuable technique. As far as this author is aware, such an experiment has yet to be tried.

REFERENCES

1. Clark, D. T. in *Handbook of X-ray and ultraviolet photoelectron spectroscopy*, D. Briggs (Ed.), Heyden, London 1977, p. 233.
2. Briggs, D. *ibid.*, p. 154
3. Seah, M. P. and Dench, W. A., *Surf. Interface Anal.*, **1** (1979), 1.
4. Roberts, R. F., Allara, D. L., Pryde, C. A., Buchanan, D. N. E. and Hobbins, N. D., *Surf Interface Anal.*, **2** (1980), 5.
5. Wagner, C. D., *Anal. Chem.*, **44** (1972), 1050.
6. Berthou, N. and Jorgensen, C. K., *Anal. Chem.*, **47** (1975), 482.
7. Clark, D. T. and Thomas, H. R., *J. Polym. Sci., Polym. Chem. Edn.*, **16** (1978), 791.
8. Seah, M. P., *Surf. Interface Anal.*, **2** (1980), 222.
9. Holm, R. and Storp, S., *Surf. Interface Anal.*, **2** (1980), 96.
10. Briggs, D. and Kendall, C. R., *Polymer*, **20** (1979), 1053.
11. Everhart, D. S. and Reilley, C. N., *Anal. Chem.*, **53** (1981), 665.
12. Everhart, D. S. and Reilley, C. N., *Surf. Interface Anal.*, **3** (1981), 126.
13. Clark, D. T., Adams, D. B., Dilks, A., Peeling, J. and Thomas, H. R., *J. Electron. Spectrosc*, **8** (1976), 51.

14. Pireaux, J. J., Riga, J., Caudano, R. and Verbist, J. J., *Proc. 1980 ACS Meeting*, Houston.
15. Clark, D. T., Feast, W. J., Musgrave, W. K. R. and Ritchie, I., *J. Polym. Sci. Polym. Chem. edn.*, **13** (1975), 857.
16. Briggs, D., Zichy, V. J. I., Brewis, D. M., Comyn, J., Dahm, R. H., Green, M. A. and Konieczko, M. B., *Surf. Interface Anal.*, **2** (1980), 107.
17. Williams, D. E. and David, L. E., in *Characterization of metal and polymer surfaces*, Vol. 2, L.–H. Lee (Ed)., Academic Press, New York, 1977, p. 53.
18. Hayes, D. A., *J. Chem. Phys.*, **61** (1974), 1455.
19. Johnson, W. T. M., *Off. Digest Oil Colour Chem. Assoc.*, **32** (1960), 1067.
20. Willis, H. A. and Zichy, V. J. I., in *Polymer surfaces*, D. T. Clark and W. J. Feast (Eds.), Wiley, London 1978, p. 287.
21. Harrick, N. J., *Internal reflection spectroscopy*, Interscience, New York 1967.
22. Wilks, P. A. Jr. in *Laboratory methods in infrared spectroscopy*, R. G. J. Miller and B. C. Stace (Eds.), Heyden, London 1972, p. 207.
23. Haslam, J., Willis, H. A. and Squirrell, D. C. M., *Identification and analysis of plastics*, Heyden, London 1980.
24. White, R. G., *Handbook of industrial infrared analysis*, Plenum, New York 1964, p. 175.
25. Tompkins, H. G., *Appl. Spectrosc*, **28** (1974), 335.
26. Griffiths, P. R. in *Laboratory methods in infrared spectroscopy*, R. G. J. Miller and B. C. Stace (Eds.), Heyden, London 1972, p. 84.
27. Chalmers, J. M., Kendall, C. R., and Briggs, D., unpublished results.
28. Hunt, C. P., Stoddart, C. T. H. and Seah, M. P., *Surf. Interface Anal.*, **3** (1981), 157.
29. Gardella, J. A. Jr. and Hercules, D. M., *Anal. Chem.*, **52** (1980), 226.
30. Briggs, D. and Wootton, A. B., in preparation.
31. Sparrow, G. R. and Mishmash, H. E. in *Quantitative surface analysis of materials*, N. S. McIntyre (Ed.), ASTM STP 643 (1978), 164.
32. Rabolt, J. F., Santo, R. and Swalen, J. D., *Appl. Spectrosc.*, **33** (1979), 549.
33. Bridoux, M. and Delhaye, M. in *Advance in infrared and Raman spectroscopy*, Vol. 2, R. J. H. Clark and R. E. Hester (Eds.), Heyden, London 1976, p. 140.
34. Shepherd, J. W. in *Advances in infrared and Raman spectroscopy*, Vol. 3, R. J. H. Clark and R. E. Hester (Eds.), Heyden, London 1977, p. 127.
35. Richter, K. and Peplinski, B., *Surf. Interface Anal.*, **2** (1980), 161.
36. Briggs, D., Rance, D. G., Kendall, C. R. and Blythe, A. R., *Polymer*, **21** (1980), 895.
37. Lindberg, B. J., *J. Electron Spectrosc.*, **5** (1974), 149.
38. Clark, D. T., Feast, W. J., Kilcast, D. and Musgrave, W. K. R., *J. Polym. Sci., Polym. Chem. Edn.*, **11** (1973), 389.

Chapter 5

OPTICAL AND ELECTRON MICROSCOPY

B. C. COPE

Leicester Polytechnic, Leicester, UK

1. INTRODUCTION

In Chapters 2–4 methods were described to study the chemistry of metallic and plastic surfaces. It is thus possible to follow the changes in surface chemistry caused by a surface treatment. In the present chapter the use of optical and electron microscopy to study the corresponding topographical changes is described. It has been noted in Chapter 1 that the topography of a surface can have an important effect on the resultant adhesion.

The principles of optical and electron microscopy are explained. The advantages and limitations of the various methods are discussed. An outline of the practical procedures involved, and solutions to some of the problems that arise, are given.

Examples are given showing how optical and electron microscopy has been used to examine adhesion problems with special emphasis on studies of surface treatments. A selection of micrographs of metals and plastics is presented and discussed.

2. OPTICAL MICROSCOPY

2.1. Principles

The human eye can, at a viewing distance of 25 cm, resolve two points if they are at least 0·1 mm apart. For any finer scale investigation, a microscope, literally a device for looking at small things, is needed. A simple hand lens, the original microscope and known seven centuries

95

ago, is still surprisingly useful to magnify up to about 10 times, but beyond this level of magnification, the shortness of focal length and small diameter of the lens become irritating and a magnification greater than 20 times is impracticable. For higher magnifications a compound microscope is necessary. This instrument, introduced early in the seventeenth century by Galileo and Keppler, consists of an objective or positive lens, nowadays usually consisting of several elements arranged to minimise aberrations, which produces an image that is viewed through and further magnified by an eyepiece lens (Fig. 1).

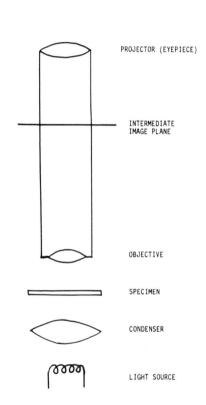

FINAL IMAGE PLANE

PROJECTOR (EYEPIECE)

INTERMEDIATE IMAGE PLANE

OBJECTIVE

SPECIMEN

CONDENSER

LIGHT SOURCE

FIG. 1. Simplified schematic diagram showing the layout of the transmission optical microscope.

The performance of a straightforward optical microscope is expressed in terms of resolving power or resolution, this being the minimum separation of two points resolvable by the viewer as separate images. The parameters controlling resolving power are the wavelength of the illumination used to form the images and the numerical aperture of the objective.

$$R = \frac{\lambda}{2NA.K} \tag{1}$$

where R is the least distance between resolvable points, λ is the wavelength of the illumination, NA is the numerical aperture and K is a constant depending on the coherence of the illumination and varying between 0·61 for white light and 1·0 for monochromatic radiation.

Numerical aperture is in turn defined by

$$NA = \eta \sin \theta \tag{2}$$

where η is the refractive index of the medium occupying the space between the specimen cover-slip and the front surface of the objective, and θ is the half-angular aperture of the objective.

It is obvious that resolution will be optimised if coherent illumination of low wavelength is used with an objective of large numerical aperture, working in a highly refractive medium. Coherence of illumination may be ensured by the use of relatively monochromatic sources and of filters passing only a very restricted spread of wavelengths.

The shortest wavelength visible to most human eyes is around 440 nm at the extreme blue end of the spectrum. The use of ultraviolet light of considerably shorter wavelength improves resolving power, but only at the expense of immediate optical visibility, the image needing to be photographically recorded before direct examination by the microscopist is possible.

Theoretically, a maximum NA of 1·0 is possible for an objective working in air, but this corresponds to a half-angular aperture of 90°, and, since this implies that the specimen is in actual contact with the front face of the objective, it is impracticable. Practical considerations on lens size and specimen–objective separation limit the half-angular aperture to about 70°, giving an NA of approximately 0·95. However, if the working medium is oil with, at 1·5, a refractive index close to that of the glass of the cover-slip, little or no internal reflection can occur leading to a condition where the numerical aperture is limited only by the small-ness of the objective and its close working distance, to an NA value of about 1·4.

Thus, by working with an oil-immersion objective and monochromatic blue light, a resolving power of $440/(2 \times 1 \times 1·4) = 157$ nm may be obtained. An objective of such resolution would produce a magnification of about 100 times, and the image produced could easily be further magnified by a $15 \times$ eyepiece to give an overall magnification of $1500 \times$. Whatever the magnification, the fundamental limit of resolving power remains, and no detail less than 157 nm will be seen. Thus magnification, after the 157 nm resolving power of the objective is enlarged to the 0·1 mm resolving power of the eye at $637 \times$, will be 'empty' and will reveal no further detail.

Oil-immersion objectives are suitable primarily for transmission micro-scopy, where the thin specimens used do not necessitate a large depth of focus, and where the illumination passes through the specimen before reaching the objective. Transmission microscopy is of very limited use in surface studies, since it is necessary to obtain thin sections for examination.

2.2. Optical surface microscopy

If it is desired to observe a surface without sectioning, it is necessary to use a microscope in which the illumination is reflected back from the surface of the specimen to the objective. This requirement, together with the need to maximise the depth of field by using a relatively small aperture objective, militates against the use both of oil-immersion tech-niques and of objectives of high resolving power. Under certain circum-stances and in experienced hands, oil-immersion objectives may be used, but the practice is not common. More frequently, objectives of modest magnification have to be used with white light and with air as the refraction medium. Consequently, substituting into eqn. (1) values of $NA = 0·65$, $\lambda = 500$ nm, $K = 0·61$ and $\eta = 1·0$, we see that the resolution is about 630 nm.

Optical surface microscopy is of considerable value in studying the topography of surfaces with surface detail appreciably larger than the resolving power of the instrument. The factor that most severely limits the usefulness of the technique is lack of depth of field (i.e. the ability to maintain a range of specimen planes in focus simultaneously). Consequently, rough surfaces cannot be observed or recorded without re-focusing between hills and valleys.

3. ELECTRON MICROSCOPY

3.1 Principles

The highly developed transmission optical microscope presses to the limits the values of numerical aperture and refractive index. The only obvious way, therefore, to improve resolution, is to decrease the wavelength of the working illumination. A change to ultraviolet rather than visible light offers some modest improvement but has concomitant disadvantages. Initial speculative experiments failed to focus X-rays, but Knoll and Ruska, applying De Broglie's discovery of the wave nature of the electron and Busch's discovery of the susceptibility of electron beams to being focused by magnetic or electrical fields, constructed in 1931 a primitive electron microscope.

When an electron beam is accelerated through a potential difference, V, the wavelength of the associated vibration is given by

$$\lambda = 1.23 \ V^{-1/2}$$

and it may be seen that when V is 20 kV, λ is 0.0086 nm, when V is 50 kV, λ is 0.0055 nm and when V is 100 kV, λ is 0.0037 nm.

Electrostatic and magnetic lenses are not able to be corrected in the same way as glass lenses and consequently the inherent spherical aberration of such lenses limits the half-angular aperture to about 1°. Even so, by applying eqn. (1), we may see that the theoretical resolving power of a 50 kV electron microscope is 0.27 nm. In practice, a new high resolution electron microscope is usually specified to maintain 1.0 nm in normal working order, a value about three orders of magnitude better than that commonly obtained with white light. Under special circumstances, resolution approaching 0.1 nm, about one interatomic spacing, is obtainable.

In addition to the superior resolving power of the electron microscope, the small aperture used in this instrument ensures a very great depth of field. The actual depth of field is dependent on many factors but may be taken as being about 1000 times the resolving power of the instrument and is very much better than that of the optical microscope.

3.2. The transmission electron microscope

The earliest type of electron microscope to achieve commercial production was the transmission electron microscope (TEM). Figure 2 shows that the optical layout in this instrument is closely analogous to that in the optical transmission microscope. Construction, of course, is

B. C. COPE

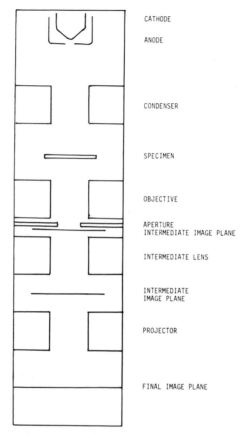

FIG. 2. Simplified schematic diagram showing the layout of the conventional transmission electron microscope (CTEM). Similarities with Fig. 1 will readily be seen.

totally different. A tungsten filament replaces the lamp of the optical microscope, magnetic lenses with an infinitely variable 'zoom' capacity replace the glass lenses and, to permit acceleration of the electron beam, the whole column is evacuated to something of the order of 10^{-4}–10^{-6} Torr with a combination of rotary and diffusion vacuum pumps.

Focusing and magnification control are effected by variation of the power of the electromagnetic lenses. The final image is projected on to a fluorescent screen for visual observation, with a diversion mechanism that enables the screen to be replaced by a photographic plate or film for the production of a permanent record. The granularity of the photographic emulsion is finer than that of the phosphor and so more detail may be shown. Magnification in the TEM is variable in steps from a few hundreds to 250,000 or more. A severe limitation of the TEM lies in the size of specimen that may be accommodated. Specimens are supported on circular grids, usually made of copper and 2·3 or 3·0 mm in diameter, but since the screen is of the order of 150 mm across, this whole area may be viewed, if at all, only at the lowest possible magnification. More seriously, the TEM is, by definition, a transmission instrument and the thickness of specimen must be limited to 10–100 nm in order to permit the passage of electrons. Most adherend samples are, of course, much thicker than this. When we are interested in bulk or internal structure, the specimen may be embedded in epoxide resin and ultramicrotomed, usually using a glass knife, to obtain suitably thin specimens. Whereas this is a standard technique for the biological specimens that constitute much of the material for TEM investigations, it is rarely applicable to surface topography studies. When surface features and topography are the object of investigation, a thin replica of the surface must usually be taken and transferred on to a copper-mesh grid acceptable to the microscope sample holder. This is a delicate procedure and one requiring both practice and a steady hand. The replica is then placed in the microscope in place of the original.

The requirements of a replica are that it shall accurately reproduce the characteristics of the surface being replicated, that it shall be thin enough to permit the transmission of electrons, and that it shall not suffer damage by the passage of the electron beam. In effect, this limits the choice to very thin films of carbon or metal laid down by vacuum evaporation or sputtering. In certain circumstances it is possible to replicate directly from the surface under examination. This method lessens the possibility of artefacts (i.e. distortion of the surface during replication, or the incorporation in the replica of features not found on the study surface as a result of operations performed during replication), but poses the problem of separation of replica and surface. The replica is too fragile to permit mechanical stripping and so the substrate must be dissolved to free the replica. It is obviously difficult to do this without destroying or damaging the replica, either by direct action of the solvent or by swelling of the substrate. Success has been achieved with polymeric substrates using solvent blends, carefully optimised to dissolve the polymer and permeate it at such a rate that swelling of the surface contiguous with the replica is restrained to such an extent that the replica is not appreciably deformed; this is Kampf's method, described by Makurak, Bojarski and Pieniazek.[1] A serious disadvantage of this method is that it is destructive of the surface being studied.

Consequently, when studying the topography of most polymers and almost all non-polymeric specimens, an indirect technique involving the production and subsequent destruction of an intermediate replica is employed. In this method, the surface under study is coated with a fine layer of low visocity solution of a low molecular weight polymer. Care must be taken to avoid solvents known or likely to alter surface morphology (e.g. acetone on polycarbonate or poly(ethylene terephthalate). After evaporation of the solvent, the polymer film is carefully stripped off and coated uniformly with carbon in a vacuum deposition chamber to provide a coherent replica. It is then obliquely shadowed with a fine grained and non-corroding metal, such as gold or gold–palladium, to reveal surface irregularities. The intermediate replica is then dissolved and the final replica 'fished out' of the solvent on a grid. In the case of low energy surfaces, those of polymers such as the fluorocarbons for instance, it may prove difficult to wet the surface with a replication solution. A way of overcoming this problem is to press the surface under study in close contact with a sheet of low molecular weight poly(methyl methacrylate), polystyrene or cellulose acetate and then prepare a replica from this.[2,3]

It is specimen preparation that most limits application of the TEM to polymer and materials science and technology in general, and surface treatment studies in particular. The instrument itself is expensive, currently a few tens of thousands of pounds, and to this must be added the cost of a vacuum coating and shadowing machine at several thousand pounds more. Coating machinery demands skilled handling and maintenance, as does all high vacuum equipment. A high level of skill and training is also needed in other stages of specimen preparation, the stripping of intermediate replicas and the mounting of the final replica on the grid for example. In the case of soluble polymers replicated directly, the dissolution of the polymer needs regular attendance and attention, it being usual to wash the polymer with solvent mixtures of controlled activity, perhaps increasing activity in a series of steps.

All of this means that although it is a fairly simple matter for the average science graduate to learn to 'drive' a TEM well enough to inspect and photograph the majority of his own samples, he is dependent on a fairly sophisticated level of technician back-up, for he is unlikely to have the time, even if possessed of the necessary patience and dexterity, to become proficient in replication techniques.

Another weakness of the TEM in surface studies is that replication of very rough or undercut surfaces is impossible. Even if possible, the very use of a replication technique may cast doubts on the validity of the resulting images. Artefacts are not always recognised for what they are, or their presence may be suspected but difficult to prove. It is certainly not unknown for artefacts to be mistaken for structural features and plausibly explained as such. Figure 3 shows just some of the ways in which artefacts may arise in surface studies and give rise to misleading representations.

3.3. The scanning electron microscope

In the mid-1960s, a type of electron microscope radically different from the TEM became available. This scanning electron microscope, SEM, made surface studies very much easier, so much easier that it could be argued that some laboratories spend too much time recording surfaces and insufficient time interpreting and explaining the resulting photographs.

Figure 4 shows the construction of the SEM. As in the case of the TEM, the optical system is contained in an evacuated tube. A tungsten filament or cathode produces electrons that are accelerated through the anode. A series of electromagnetic lenses produce a spot, a reduced image of the cathode on the specimen plane. This electron beam is scanned over the specimen, which must be electrically conducting, in a raster pattern. A large number of imaging systems have been developed based on such effects as cathodoluminescence, induced EMFs, and X-ray emission but the most widely used techniques are based on the collection of electrons.

In the emissive mode of operation, the most common, secondary electrons knocked out of the sample surface by the incident beam are used to create a visual image, whilst in the reflective mode, backscattered primary electrons are utilised.

The SEM, being essentially a scanning instrument, does not create and display a steady image, as in the TEM, but one in which the scanning spot of light, drawing out the raster pattern on the display cathode ray screen, must always intrude somewhat. This effect may be minimised by using TV scan rates on the display, but only at the cost of a high noise-to-signal ratio. It is common practice to take a preliminary look over the specimen at TV scan rate, then to 'home-in' on areas of interest and

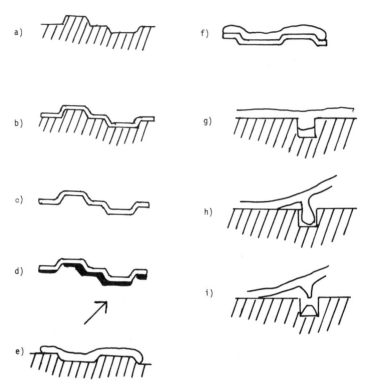

FIG. 3. Replication techniques for CTEM and artefacts that may be introduced.
(a)–(d) Single stage replication. (a) The surface to be replicated. (b) Uniform
carbon layer evaporated on to surface. (c) Substrate dissolved to free carbon
replica. (d) Replica shadowed with metal (direction of shadowing is shown by
arrow). Shadowing may be applied to either side of the replica.
(e)–(f) Two stage replication. (e) An intermediate polymer replica is taken from
the substrate and then mechanically stripped from it. (f) Intermediate replica
carbon coated. Subsequent steps are identical with (c) and (d).
   Artefacts may typically be introduced during production and stripping of the
intermediate replica. (g) Replicating liquid does not wet out hollow in surface. (h)
Replica is strained and deformed during stripping. (i) Replica fractures on
stripping.

examine these in more detail and with greater clarity using a slower scan
rate, perhaps 0·1–2 seconds per frame, on a high persistence screen and
finally to use a very slow scan rate, perhaps one frame per 100 seconds,
on a low persistence screen to produce the best image for photography.
Photography in the SEM is usually carried out with a conventional

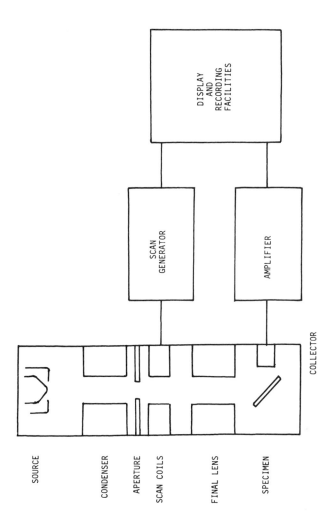

FIG. 4. Simplified schematic diagram of scanning electron microscope (SEM).

35 mm camera, with consequent convenience in processing and negative storage.

The first advantage of the SEM over the TEM lies in its ability to accommodate relatively large specimens, perhaps 10 mm square. This specimen requires much less preparation for the SEM than for the TEM. In the SEM, specimens are adhesively bonded to a small aluminium 'stub', and, if they are of conducting material permitting the passage of secondary electrons, require no further treatment. Non-conducting specimens need the deposition of a thin layer of gold or gold–palladium ($\sim$ 20 nm) to provide a conducting surface, unless very low accelerating voltages can be used, as by Brown and Swift[4] in their study of cosmetic treatments applied to the hair. Most instruments allow the specimen to be tilted or rotated to permit examination of particular features, and in some, heating, cooling and mechanical testing may be performed.

Staining and etching, as practised in transmission optical microscopy and sometimes in transmission electron microscopy, are generally not applicable in scanning electron microscopy, but certain workers have developed specialised techniques. Hawkins and Lewis, for example, have intensively investigated the etching and polishing of polycarbonates, and have published formulae for both etching (i.e. accentuating topography, particularly to show up crazing or cracking) and polishing (i.e. reducing topographical features and thinning uniformly).[5] Atkinson et al.,[6] Kato[7] and Kambour[8] have also worked with heavy dopants, such as osmium tetroxide, silver and iodine to show up small cracks.

The depth of field of the SEM is extremely good, exceeding even that of the TEM, but not, of course, infinite. A strong stereoscopic effect is also produced. The range of magnification commonly provided is from perhaps 20 or 50 up to as much as 50 000 plus.

If the chief advantages of the SEM over the TEM are ease of specimen preparation and confidence in verisimilitude of image, then the chief disadvantage lies in the comparatively poor resolution. The finite size of the electron beam 'spot' and the impossibility of providing a perfectly stable power supply to the lenses, effectively place a limit on the resolution of the SEM. Very high quality research instruments may, under good conditions, maintain a resolving power of about 5 nm, whereas a small instrument in a routine inspection laboratory might maintain 10 nm.

A second defect is the susceptibility of organic materials to damage from irradiation by the electron beam. Under beams of high energy, the sample may deform during inspection, leading to erroneous conclusions.

3.3.1. Modes of operation of the SEM

In the SEM, the collection of information from reflected primary electrons, emitted secondary electrons or other sources induced or influenced by the interaction of the electron beam with the specimen, is purely electronic; no visible image is formed before the final cathode ray tube display, so signal processing to a high degree is possible.

The SEM's normal mode of operation is based on the scanning, in a rectilinear raster pattern, over the specimen surface of the primary electron beam and the collection of the emitted secondary electrons, the so-called emissive mode. The degree of magnification is controlled by variation of the ratio of the raster on the specimen to that on the display CRT. The emissive mode gives great topographical contrast, but usually does not give compositional information. Contrast is also produced partly by differences in specimen conductivity, less conducting areas giving higher yields of secondaries. Electric and magnetic field contrast studies are possible in the emissive mode, but are carried out only on ferromagnetic materials.

Voltage contrast imaging in the emissive mode is useful for investigations on semiconductors. These depend on topical variations in the potential difference between the specimen and its surroundings.

Switching the bias on the collector grid from the large positive voltage emissive mode to a small negative voltage, filters out low energy secondaries and puts the instrument into the reflective mode, where the image is formed from backscattered primaries. The reflective mode again gives topographical contrast, but is more sensitive than is the emissive mode to compositional contrast produced by atomic number variation. The differences in resolution obtained via the reflective and emissive modes arise from the different depths relative to the surface at which the image-forming electrons originate. The reflective mode shows less detail in deep pits than does the emissive mode.

Other imaging modes are of less interest to adhesives and coatings technologists. For example, electron channelling patterns (ECP) give crystallographic data, whilst induced conductivity imaging is suitable for investigating semiconductors. Absorbed current imaging utilises the picoamp currents that flow if the specimen is earthed, and whose magnitude is composition-sensitive, to give a picture in which topographical detail is sacrificed to enhanced compositional information. Cathodoluminescence is also a mode of operation that concentrates on composition, particularly in relation to crystal impurities and semiconductors.

3.3.2. *Microanalysis in the SEM*

The SEM may be constructed or adapted to operate as an analytical instrument by using the electron beam to excite X-rays or Auger electrons. These techniques are well known in their own right, and electron probe microanalysers and Auger spectrometers are also well known, even if expensive and hence relatively uncommon.

From the point of view of the adhesives or coatings technologist, the ability of the SEM to be easily switched from producing a topographical image to investigating the composition of the specimen, is particularly useful.

In X-ray microanalysis, a fairly high voltage beam is used to excite X-rays in a volume of about 1 μm^3 of the sample (which must of course be uncoated). The emitted X-rays are analysed in one of two ways. In the wavelength dispersive (WD) method a crystal spectrometer is used. The X-ray beam strikes the crystal, and only rays fulfilling the Bragg's Law condition reach the counter, where they impart a voltage pulse proportional to the energy of the X-ray photon. Several crystals are needed to cover the complete emitted X-ray wavelength range of 0·1–10 nm. Resolution in the WD system is good, but the maintenance of stringent standards of alignment of the total assembly is critical.

In energy dispersive microanalysis (EDAX), the entire emitted X-ray spectrum is simultaneously received, and the detector differentiates by energy (and hence by wavelength). The CRT display may be set to give a direct visualisation of the X-ray spectrum, intensity being plotted against photon energy. This may be labelled as relative abundance against atomic number of originating nucleus to show specimen composition directly. EDAX resolution is an order of magnitude less than WD and low energy X-rays cannot be detected at all, leading to inability to analyse for elements lighter than atomic number 10. Against this, the robustness of the apparatus and its rapidity and ease of operation and maintenance have made EDAX much more popular than WD systems, and many SEMs are fitted with the device, either as original equipment or as a retrospective modification. Some quantitative investigations may be made or specimens may be scanned for X-rays of given energy to give composition maps for particular elements. Specimens must be uncoated, of course, and since carbon, hydrogen and oxygen are not detectable, the usefulness of the technique to study surface modification of polymers and oxide formation on metals is limited.

4. RECENT DEVELOPMENTS

Conventional TEMs work with an accelerating voltage in the order of 100 kV. An increase in voltage would permit better resolution and the penetration of thicker samples at the possible cost of increased beam damage. Consequently, instruments operating at 1 MV up even to 10 MV are in use or under construction, but they have little obvious application to polymer surface studies.

The scanning transmission electron microscope, STEM, makes more efficient use of transmitted electrons than does the TEM and when operating at high voltages can produce good resolution from a comparatively thick specimen. Initially modified from existing TEMs, such instruments are now purpose built and in use with both inorganic and biological specimens. Polymer applications have been slower to develop and little use of the technique has yet been made by adhesives technologists. Such instruments may also produce good resolution at a comparatively low voltage and are able to produce selected area diffraction patterns that can be made to yield structural information. Also, they possess the versatility in imaging and image processing of the SEM.

5. PRACTICAL APPLICATIONS OF MICROSCOPY IN ADHESION PROBLEMS

5.1. Transmission optical microscopy

Transmission optical microscopy is a technique that demands that samples be presented in the form of transparently thin sections. Application to surface and adhesion studies thus demands sectioning of the subject material, be it an adherend, surface treated or 'as received', or a complete assembly of adherends and adhesive or substrate plus surface coating. Obviously opaque materials such as metals and minerals are unable to be sectioned and the utility of the technique is limited to organic specimens.

Biological specimens are rarely of interest to the adhesion scientist but it is in this area that embedding, sectioning and staining techniques have been brought to a high degree of excellence. Most specimens for adhesion studies are likely to be polymers, either natural or synthetic, and these can often be difficult to section. Being tough, resilient and moderately flexible they deform rather than cleave under the microtome blade. In order to harden polymers for sectioning, the technique of freezing or, more accurately, cooling below the T_g has become the norm under the name 'cryomicrotomy'. The history of cryomicrotomy can be traced back as far as 1920 when Depew and Ruby[9] first successfully sectioned rubbers, whilst the use of glass blade microtomes for such sectioning was introduced by Latta and Hartmann[10] 30 years later. Whereas cryomicrotomy is applicable to polymers in general, a hardening technique, whose applicability is limited to elastomers, is Roningen's method,[11] in which molten sulphur is used to convert the elastomer present to rigid ebonite.

Microtomy is not an easy technique and knife-marking and other problems are not uncommon. Consequently, some workers have sought to gradually grind or machine thin samples from bulk specimens, using techniques adapted from metallographic and petrographic specimen preparation procedures. Such specimens are usually mounted in epoxide resin before size reduction. A procedure of general applicability is described by Holik et al.[12]

Specialised techniques peculiar to one type of material or joint have been evolved by many workers and are often published in the relevant trade journals rather than in the literature of microscopy.

A typical adhesion study, based on transmission optical microscopy, is that of Smith[13] on the bondability of rubber exposed to surface oxidation and degradation. Haines examined adhesion to surface treated leather[14] by using dyes which coloured the adhesive but not the substrate, whereas Sharpe has examined the interphase region in polyethylene and its effect on adhesion,[15] and James has used transmission optical microscopy to study rubber to tyre cord adhesion.[16,17]

5.2. Surface optical microscopy

Little or no specimen preparation is needed with surface optical microscopy, although sometimes a metal coating is evaporated on to reduce reflection from the intensely white surfaces of highly crystalline polymers such as PTFE (D. J. Barker, pers. comm.). The technique is limited in its scope by the relatively low resolution and, more important, limited depth of focus that it offers. But these defects may be outweighed in some circumstances by the ease, rapidity and cheapness that it possesses. A useful area of application where these benefits are exploited is in the study of corrosion at the adherend–adhesive interface as examined, for example, by Schultz et al.,[18] in their study of the surface pretreatment of steel and its effects on the peel strength of electrodeposited, epoxide-based, protective coatings.

An interesting comparison of the appearance of the same aluminium surface, as viewed by surface optical microscopy and emissive mode scanning electron microscopy, is given by Bochynski.[19]

Useful information can sometimes be obtained by sectioning a joint, preparing the cut surface by metallographic polishing and then examining it in the surface microscope.[20]

5.3. Transmission electron microscopy

Like transmission optical microscopy, TEM (or CTEM, conventional transmission electron microscopy), demands a very thin specimen and is sensitive to compositional differences as reflected by density; hence it is particularly suited to investigations of sections of bonds. However, the fact that even thinner specimens are needed than with transmission optical microscopy, means that sectioning is even more difficult. Cryomicrotomy and, where appropriate, Roningen's method, are widely used. Smith[13] used both in examining rubber–rubber bonding.

The literature of cryomicrotomy is fairly extensive and often specialised, but general practical details can be found in Cobbold and Mendelson,[21] Pullen et al.,[22] and Attenburrow and Lewis.[23]

The difficulty of sectioning and the susceptibility of polymers to beam damage have both served to inhibit TEM studies on sections, but one might note Anderson et al. on the adhesion of dental prosthetics,[24] and Aspbury and Phillips on epoxide adhesives and the presence therein of nodular structures and their influence on post-cure properties.[25]

Surface studies using TEM (both CTEM and STEM) are much more tedious than those employing SEM, but workers such as Menter[26] have been able to reveal surface defects and topographical features well beyond the resolving power of SEM.

The mode of formation and morphology of the oxide layers produced on aluminium by etching and anodising treatments have been investigated by Bijlmer,[27–29] Pattnaik and Meakin[30,31] and Venables et al.[32,33]

Other studies are typified by Hawthorne and Teghtsoonian[34] on carbon fibre surfaces, Ketkar and Keller[35] on adhesion to mineral surfaces and Van Gunst et al.[36] on the tack of ethylene–propylene elastomers, where the authors identified a second phase, responsible for the existence of tack, by using TEM on replicas.

5.4. Scanning electron microscopy

The development of SEM, with its ease of specimen preparation and examination, reduced the time taken to examine a surface from hours to minutes, and, not surprisingly, inspired a large number of novel investigations. Adherend surfaces and fractured joints were early candidates for examinations, by Lee et al.[37,38] amongst others, who examined both adherend surfaces and failed joints to clarify failure mechanisms.

In those early days there was a tendency to use the technique as an end in itself rather than as a tool; more attention was paid to recording surfaces than interpreting the surfaces thus portrayed. In due course, the novelty faded and we now find SEM applied, often in conjunction with other techniques, to solving adhesion problems. Most investigations are not published for a wider audience than that to which an internal laboratory report circulates, but the research literature nevertheless contains many examples illustrating the usefulness of SEM techniques in surface studies. These papers are often not easy to trace, as, even when the microscopy is an important aspect of the investigation, no reference to microscopy may appear in the title or in the keyword abstract.

Some important studies on the surface treatment of titanium have been made by Wake, Allen and Alsalim.[39,40] The use of copper as a substrate and the effect of its morphology on adhesion has been extensively reported by Evans, Packham and their collaborators[41-44] and by Wang and Vazirani.[45] Evans and Packham have also worked on zinc and steel.[44] Steel has also been examined by Baker[46] and Schultz et al.,[18] and brass by Haemers[47] in the course of a study of the adhesion of rubber to steel tyre-cord.

However, the most extensively investigated metal substrate in adhesion studies is aluminium, some of the more interesting observations emanating from Thrall[48] in a series of publications presenting results from the USAF Primary Adhesively Bonded Structure Technology (PABST) programme, in which a large part of a heavy transport aircraft was constructed using adhesive bonding, and was extensively tested, Wyatt et al.,[49] Bijlmer[28,29,50] and Bochynski.[19] Danforth and Sunderland[51] used SEM studies to illustrate the contamination and damage caused to treated aluminium by normal handling and storage procedures, and related this to poor bonding performance and premature bond failure.

A considerable effort has been devoted to studies of adhesion to carbon fibres, and several publications have included SEM micrographs. Good examples are found in work by Kaelble and Dynes,[52] and Dauksys.[53]

An unusual substrate investigated by Batchelor et al.[54] is concrete in the context of adhesion of polyurethane elastomer-based waterproofing membranes to concrete bridge decks. Glass has been examined by Vera et al.[67] using both optical and electron microscopy.

Electron microscopic investigations of polymers are more often directed towards morphology studies than adhesion; Purz and Schulz[55] have recently reviewed work in this field. Mention might, however, be made of the work on polypropylene by Garnish and Haskins,[56] who examined the physical roughening of surfaces by treatments designed to improve adhesion, such as chromic acid pickling and vapour etching; on polyurethane adhesives and rubber by Pettit and Carter;[57] and on polymeric dental fillers by Lee et al.[58,59] ABS has received considerable attention, particularly from Atkinson et al.,[6] Bucknall[60] and Gentle.[61]

Deakyne and Whitwell[62] have used the SEM in investigating the bonding of non-woven fabrics with polymeric adhesives, whilst knitted fabrics were similarly examined by Andrews et al.[63] The surface treatment of polyethylene has been evaluated by Blais et al.[64] and Shields.[65]

Another aspect of adhesion that has been investigated using the SEM is the incorporation of carbon black dispersions in the glue line, a matter investigated by Corish et al.[66] and found by them to promote adhesion of dissimilar rubbers that were otherwise difficult to bond.

6. THE PLACE OF MICROSCOPY IN THE ADHESIVES LABORATORY

Large organisations with centralised corporate research and development facilities may be able to afford the investment in instruments and specialised manpower that is demanded by a complete optical and electron microscopy service, but many producers of adhesives and coatings and many more users of these products are not able to dedicate this much effort. There is, however, a fairly clearly defined gradation in terms of information gained per unit of finance and effort expended. We may take it that the transmission optical microscope is so limited or, perhaps, so specialised in its capability to assist the adhesives technologist that its provision is either an extreme luxury or a necessity.

The surface optical microscope is relatively cheap and robust; it requires no specialised services and the training needed to enable general scientific personnel to operate it, and even maintain it, is small. The amount of information derived from its use is relatively small, but specimen preparation is very quick and easy. One could reasonably conclude that a *simple* surface optical microscope should be available in every adhesives laboratory.

When we turn to SEM, there is a quantum jump in capital cost from one or two thousand pounds to tens of thousands. The scanning electron microscope requires an electrically and magnetically stable environment, cooling water, specialised high vacuum and electronics maintenance, and, if non-metallic specimens are to be examined, vacuum metallisation or sputtering facilities (costing several thousand pounds). A simple bench-top instrument, operating only in the emissive mode and not fitted with analytical facilities, is within the financial reach and operating capabilities of many adhesives producers and many adhesives users. Such an instrument can be as much part of the laboratory's general user equipment as, say, an infrared spectrophotometer, and will almost certainly achieve a high utilisation factor.

A more sophisticated scanning electron microscope with alternative operating modes and, perhaps, analytical facilities, needs a dedicated technician both to maintain it at its specification capability and to realise this capability in daily use. This type of instrument, in fact, crosses the boundary between the freely available laboratory equipment and the specialised service.

A useful solution for companies requiring the use of such an instrument, and, in particular, needing the improvement in performance that it offers over the bench-top scanning electron microscope, but unable to justify the capital outlay or the dedication of personnel, is to hire time on an hourly basis on a large microscope in the laboratory of an academic institution or research or consultancy organisation. The rates in such institutions are often relatively modest, and the only real disadvantage of using such equipment is the difficulty of emergency or priority access.

CTEM, and even more conspicuously, STEM, are the province of the specialist and few adhesives producers could begin to justify the provision of such a service. The users of adhesives who require the precision of data made available from these techniques are usually best advised to hire time (and assistance) on institutional instruments, but in some areas, such as electronics, the customer companies are sufficiently large and technologically aware to be able to locate a TEM facility in their central research and development operation.

7. CONCLUSIONS

In understanding the changes brought about in substrate surfaces by treatment designed to improve adhesion, it is important, but often forgotten, to study not only chemical modification but also changes in topography and morphology. For this purpose, microscopy, either optical or electron, is usually the most appropriate technique.

Scanning electron microscopy offers a useful combination of advantages for most applications. Sample preparation is quick and simple. High resolution and high magnification are achieved in combination with an extremely large depth of focus; elemental analysis of suitable substrates is possible. Photographic recording is simple and cheap. The instrument may be operated, unless optimum performance in specialised functions is required, by the adhesives scientist or technologist himself, rather than by a specialist microscopist.

These advantages have led to the widespread use of the SEM in industry and in many academic projects.

Where resolution is all important and where the concomitant disadvantages of lengthy and difficult specimen preparation (with a significant danger of artefact product) are tolerable, transmission electron microscopy is the preferred technique. The use of TEM in industry is relatively uncommon and normally limited to the more fundamental studies carried out by the larger polymer manufacturers and users of adhesives in the high technology areas. Even in academic programmes SEM is preferred to TEM where possible, but in fundamental investigations TEM is often found indispensible.

Optical microscopy is frequently underestimated. Although resolution and depth of focus are severely restricted, they are often sufficient for technological projects. The advantages of optical over electron microscopy lie in low cost, freedom from maintenence demands, durability and simple preparation of specimens.

ACKNOWLEDGEMENTS

The author thanks Mr. D. W. Bazeley, Mr. C. Warrington, Mr. J. D. Gribbin and Miss G. Mander all of Leicester Polytechnic, for their help in preparing the unattributed figures.

APPENDIX: A SELECTION OF TYPICAL MICROGRAPHS

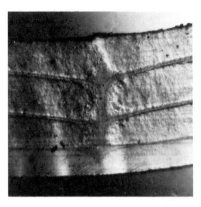

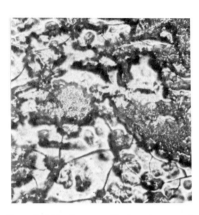

FIG. A1 Transmission optical micrograph of cut section showing interfaces between strands of polypropylene laminated into pipe. Magnification × 30.

FIG. A2 Surface optical micrograph showing corrosion of a copper–beryllium alloy. Note its limited depth of focus. Magnification × 200.

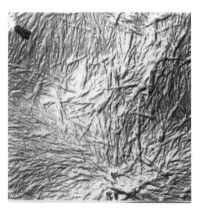

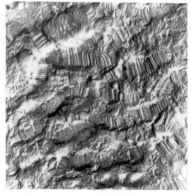

FIG. A3 Transmission electron micrograph of direct replica of polycarbonate surface showing crystallisation as a result of the use of chlorinated solvents in adhesive cements. Magnification × 13 000.

FIG. A4 Replica electron micrograph of PTFE surface treated with sodium napthalenide to promote adhesion. Magnification × 40 000. Extensive crystalline regions are visible. It is thought that the process of stripping the cellulose acetate replica from the PTFE mechanically removes some form of PTFE, leading to a false impression of the surface being recorded on the final carbon replica. This is an unusual example of an artefact.

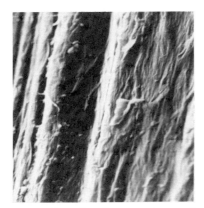

FIG. A5 Scanning electron micrograph of extruded polypropylene as produced. Magnification × 1250.

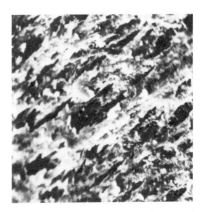

FIG. A6 Scanning electron micrograph of polypropylene in Fig. A5 etched in freshly prepared chromic acid. Magnification × 1250.

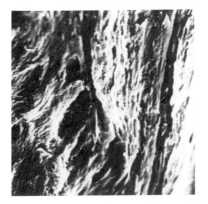

FIG. A7 Scanning electron micrograph of polypropylene in Fig. A5 etched in aged chromic acid. Magnification × 1250.

FIG. A8 SEM of titanium etched in hydrofluoric acid. Magnification × 2500. (Allen and Alsalim, reference 39).

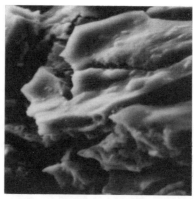

FIG. A9   SEM of titanium surface coated with titanium and ruthenium oxides for use in electrochemical preparations (magnification × 5000). This micrograph demonstrates the extremely good depth of focus of the SEM and shows a rough and undercut surface not susceptible to reliable replication.

REFERENCES

1. Makurak, I., Bojarski, J. and Pieniazek, J., *Polymery*, **13** (1968), 341.
2. Bunn, C. W., Cobbold, A. J. and Palmer R. P., *J. Polym. Sci.*, **28** (1958), 365.
3. Speerschneider, C. J. and Li, C. H., *J. Appl. Phys.*, **37** (1962), 1871.
4. Brown, A. C. and Swift, J. A., *Proc. 5th Eur. Congr. Electron Microsc.*, (1972), 386.
5. Hawkins, P. and Lewis, P., Paper read at S.C.I. Conference, *Toughness in Homogenous Polymers*, London, 31 October 1978.
6. Atkinson, E. B., Brooks, P. R., Lewis, T. D., Smith, R. R. and White, K. A., *Trans. J. Plast. Inst.*, (1967), 550.
7. K. Kato, *J. Electromicrosc.*, **19**(1) (1970), 1.
8. Kambour, R. P., *Polymer*, **5**(3) (1964), 143.
9. Depew, H. A. and Ruby, I. R., *Ind. Eng. Chem.*, *Anal. edn.*, **2** (1920), 1156.
10. Latta, H. and Hartmann, J. F., *Proc. Exp. Biol. Med.*, **74** (1950), 436.
11. Roningen, F. H., *Ind. Eng. Chem.*, *Anal. edn.*, **5** (1933), 251.
12. Holik, A. S., Kambour, R. P., Fink, D. G. and Hobbs, S. Y., *Microstruct Sci.*, **7** (1979), 357.
13. Smith, R. W., in *Adhesion science and technology*, Vol. 9 A, L. H. Lee (Ed.), Plenum, New York 1975, p. 289.
14. Haines, B., in *Aspects of adhesion*—3, D. J. Alner (ED.), University of London Press, London 1967, p. 40.
15. Sharpe, L. H., *J. Adhes.*, **4** (1972), 51.
16. James, D. I. and Wake, W. C., *Trans. Inst. Rubb. Ind.*, **39** (1963), T103.
17. James, D. I., Norman, R. H. and Stone, M. H., *Plast. Polym.*, **36** (1968), 21.
18. Schultz, J., Sehgal, K. C. and Shanahan, M.E.R. in *Adhesion—1*, K. W. Allen (Ed.), Applied Science, London 1978, p. 269.

19. Bochynski, Z., *Proc. 5th Eur. Congr. Electron Microsc.*, (1972), 674.
20. Baier, S. W. in *Aspects of adhesion—1*, D. J. Alner (Ed.), University of London Press, London 1965, p. 108.
21. Cobbold, A. J. and Mendelson, A. E., *Science Tools*, **18** (1971), 1.
22. Pullen, S. F., Fleet, E. C., Meyer, D. E. and Thomas, K., *Proc. 5th Eur. Congr. Electron Microsc.*, (1972), 258.
23. Attenburrow, G. E. and Lewis, P. R., *ibid*, 256.
24. Anderson, G. P., Koblitz, F. E., Glenn, J. F. and Devries, K. L., *J. Adhes.*, **9** (1978), 213.
25. Aspbury, P. J. and Phillips, M., in *Adhesion—1*, K. W. Allen (Ed.), Applied Science, London 1977, p. 223.
26. Menter, J. W., *J. Inst. Metals*, **81** (1952), 163.
27. Bijlmer, P. F. A. and Schliekelmann, R. J., *SAMPE Quart.* (13 October 1973).
28. Bijlmer, P. F. A. in *Adhesion—2*, K. W. Allen (Ed.), Applied Science, London 1978, p. 45.
29. Bijlmer, P. F. A., *J. Adhes.*, **5** (1973), 319.
30. Pattnaik, A. and Meakin, J. D., *J. Appl. Polym. Sci.*, *Appl. Polym. Symp.*, **32** (1977), 145.
31. Pattnaik, A. and Meakin, J. D., *Picatinny Arsenal Tech. Report No.* 4699 (July 1974).
32. Venables, J. D., McNamara, D. K., Chen, J. M., Sun, T. S. and Hopping, S., *Nat. SAMPE Tech. Conf.*, **10** (1978), 362.
33. Chen, J. M., Sun, T. S. and Venables, J. D., *Nat. SAMPE Symp. Exhib.*, **22** (1978), 25.
34. Hawthorne, H. M. and Teghtsoonian, E., *J. Adhes.*, **6** (1974), 85.
35. Ketkar, A. B. and Keller, D. V., *J. Adhes.*, **7** (1975), 235.
36. Van Gunst, C. A., Paulen, H. J. G. and Wolters, E., in *Adhesion—1*, K. W. Allen (Ed.), Applied Science, London, 1977, p. 85.
37. Lee, H. L. and Stoffey, D. G., *Chem. Eng. News* (23 September 1968).
38. Cagle, C. V. and Lee, H. L., *Adhesives Age*, **714** (1971), 40.
39. Allen, K. W. and Alsalim, H. S., *J. Adhes.*, **6** (1974), 229.
40. Allen, K. W., Alsalim, H. S. and Wake, W. C., *J. Adhes.*, **6** (1974), 153.
41. Adam, T., Evans, J. R. G. and Packham, D. E., *J. Adhes.*, **10** (1980), 279.
42. Evans, J. R. G. and Packham, D. E. in *Adhesion—1*, K. W. Allen (Ed.), Applied Science, London 1977, p. 297.
43. Evans, J. R. G. and Packham, D. E., *J. Adhes.*, **10** (1979), 39.
44. Evans, J. R. G. and Packham, D. E., *J. Adhes.*, **10** (1979), 177.
45. Wang, T. T. and Vazirani, H. N., *J. Adhes.*, **4** (1972), 353.
46. Baker, F. S., *J. Adhes.*, **10** (1979), 107.
47. Haemers, G., in *Adhesion—4*, K. W. Allen (Ed.), Applied Science, London 1980, p. 175.
48. Thrall, E. W., in *Adhesion—4*, K. W. Allen (Ed.), Applied Science, London 1980, p. 1.
49. Wyatt, D. M., Gray, R. C., Carver, J. C., Hercules, D. M. and Masters, L. W., *Appl. Spectrosc.*, **28** (1974), 439.
50. Bijlmer, P. F. A., *Proc. 4th Int. Symp. Contam. Control* (1978), 247.
51. Danforth, M. A. and Sunderland, R. J., *J. Appl. Polym. Sci.*, *Appl. Polym. Symp.*, **32** (1977), 201.

52. Kaelble, D. H. and Dynes, P. J., *J. Adhes.*, **6** (1974), 239.
53. Dauksys, R. J., *J. Adhes.*, **5** (1973), 211.
54. Batchelor, J., Robinson, M. and Lambropoulos, V. L. in *Adhesion—1*, K. W. Allen (Ed.), Applied Science, London 1977, p. 53.
55. Purz, H. J. and Schulz, E., *Acta Polymerica*, **30**(7) (1979), 377.
56. Garnish, E. W. and Haskins, C. G., in *Aspects of adhesion—5*, D. J. Alner (Ed.), University of London Press, London 1969, p. 259.
57. Pettit, D. and Carter, A. R., *J. Adhes.*, **5** (1973), 333.
58. Lee, H., Swartz, M. L. and Stoffey, D. G., *Appl. Polym. Symp.*, **16** (1971), 1.
59. Lee, H., Swartz, M. L. and Stoffey, D. G., *ACS Div. Org. Coatings, Plast. Chem. Pap.*, **30** (1970), 243.
60. Bucknall, C. B. and Drinkwater, I. C., *Polymer*, **15** (1974), 254.
61. Gentle, D. F. in *Aspects of adhesion—5*, D. J. Alner (Ed.), University of London Press, London 1969, p. 142.
62. Deakyne, C. K. and Whitwell, J. C., *J. Adhes.*, **8** (1977), 275.
63. Andrews, B. A., Joynes, W. R., Gautreaux, G. A. and Frick, J. G., *Microscope*, **21** (1973), 161.
64. Blais, P., Carlsson, D. J., Csullog, G. W. and Wiles, D. M., *J. Coll. Interface Sci.*, **47** (1974), 636.
65. Shields, J., *SIRA Report* R500 (1972).
66. Corish, P. J., Osmant, T. H. and Clarke, D. I. in *Adhesion—4*, K. W. Allen (Ed.), Applied Science, London 1980, p. 199.
67. Vera, R., Baer, E. and Fort, T., *J. Adhes.*, **6** (1974), 357.

Chapter 6

THERMODYNAMICS OF WETTING: FROM ITS MOLECULAR BASIS TO TECHNOLOGICAL APPLICATION

D. G. RANCE

ICI Petrochemicals and Plastics Division, Welwyn Garden City, UK

1. INTRODUCTION

This chapter reviews molecular interactions which occur at interfaces. It shows that the wetting or non-wetting of solid surfaces by liquids can be described in terms of thermodynamic parameters, such as surface and interfacial free energy, which characterise the interacting materials. Surfaces are often classified according to whether they are high energy surfaces or low energy surfaces. Generally, high energy surfaces are those which have a surface free energy > 100 mJm^{-2} and include inorganic solids, glasses and metals. On the other hand, low energy surfaces have a surface free energy < 100 mJm^{-2} and include all organic liquids, waxes and organic polymeric solids. This immediately provides a distinction between polymers and metals; hence they can be expected to interact differently with liquids with which they are brought into contact.

A frequently used measurement for obtaining information about surfaces is the contact angle which a liquid makes with a solid. Since a liquid makes contact with the outermost molecular layer of a surface, contact angles are more sensitive to chemical and structural changes which occur in a surface than other techniques for surface analysis, e.g. MIR and XPS. Contact angle measurements are therefore useful for monitoring surface changes which are produced as a result of some pretreatment.

A review of the methods of contact angle measurement is included in this chapter, together with information on how the contact angle is modified by a surface which is either rough or chemically heterogeneous. The use of contact angle measurements to calculate the surface free energy of solids is outlined, but it should be remembered that thermo-dynamic parameters can only be determined with confidence for surfaces which are both smooth and homogeneous and under conditions where a state of equilibrium with the environment has been reached. However, the main limitation in the determination of surface free energy of solids arises from the nature of the approximations to thermodynamic equations which are often made.

Many analyses of interfacial interactions overlook the very specific and relatively strong interactions arising from certain chemical groups which have some hydrogen bonding capability. Hence, the study of hydrogen bonding interactions (considered as a type of acid–base inter-action) by contact angle measurements represents an important deve-lopment in the understanding of the nature and strength of interfacial interactions.

The literature abounds with different correlations between measured adhesive strength and what purport to be thermodynamic parameters of interfaces but they have not all been considered here. This chapter highlights important developments in the study of the physical chemistry of interfaces; for more detailed information on the material covered, the reader is referred to the most recent and relevant reviews.

2. MOLECULAR INTERACTIONS ACROSS INTERFACES

The surface tension of a solid or liquid has been defined by Fowkes[1,2] as a measure of the attractive forces which act in the surface layers of the material and interfacial tension as a measure of the attractive forces between molecules of the materials which have been brought into intimate molecular contact. There are three types of attractive forces between atoms or molecules which may operate at interfaces.

(a) Primary bonding forces, such as covalent or electrostatic bonds, where the binding energies between atoms (40–400 kJ/mole) pro-vide the strongest bonds.
(b) Secondary bonding forces, such as van der Waals forces, the weakest bonds with bond energy of 4–8 kJ/mole.
(c) Hydrogen bond forces, which produce bonds intermediate in bind-ing energy at 8–35 kJ/mole.

Primary bonding forces will result in very strong interfacial adhesion. This can be produced by interfacial cross-linking reactions between polymers, or between polymer and metal where the metal has been treated with a coupling agent such as a siloxane-containing terminal reactive group. However, only secondary bonding forces and hydrogen bond forces are relevant when considering the adsorption mechanism of adhesion.

Secondary forces can be subdivided, according to the nature of the interaction. London dispersion forces describe the interaction between completely non-polar atoms or molecules. Attraction between two atoms or molecules results if an instantaneous fluctuation in one atom polarises electrons in an adjacent atom or molecule. Keesom forces act between molecules possessing permanent dipoles; Debye forces describe interactions between dipoles and induced dipoles. The most important of these secondary forces are London forces and it will be shown later that other secondary forces provide a negligible contribution to the surface free energy of most solids. The literature refers to London forces as dispersion forces, a term which Good[3,4] has suggested is inappropriate. The name arises from optical dispersion, which provides a relationship between the refractive index of a material and the frequency of light at which the measurement was made. From the dispersion, both the electronic polarisability of the molecule and a characteristic absorption frequency in the ultraviolet region can be calculated, parameters which were used by London to calculate the attractive forces between molecules.

The attraction between simple molecules was extended to surfaces by Hamaker,[5] who summed the attraction between individual oscillators constituting two surfaces. This so-called microscopic approach related the force of attraction between two surfaces considered as parallel plates to the inverse cube of their distance of separation. The constant of proportionality, the Hamaker constant, is a function of the magnitude of London dispersion forces which act in each surface and the intervening gap. Gregory[6] and Visser[7] have reviewed Hamaker constant measurements for a number of materials including polymers, metals and metal oxides. As the separation between surfaces increases, a relativistic correction is needed to account for the finite time required to propagate electromagnetic radiation across the gap. Consequently, the attraction is reduced so that when retardation begins to occur, the attractive force between surfaces falls more rapidly than the inverse cube of the distance of separation. Tabor and Winterton[8] have shown experimentally that the changeover from unretarded to retarded van der Waals forces for crossed mica hemi-cylinders across an air gap occurs at a separation of about 20 nm.

Where dispersion forces are the most important in controlling the wetting of surfaces, the microscopic theory has been used as shown in section 7.1 to relate intermolecular forces to thermodynamic parameters of a surface. While this may be a good approximation for non-polar surfaces which interact only as a result of electronic perturbations, it is not a good approximation for all materials. In recent years the macroscopic approach suggested by Lifshitz[9,10] has been widely accepted as a more realistic method of calculating the force or free energy of attraction between macroscopic bodies. The electrodynamic theory of Lifshitz is based on the quantum field approach, which allows for many-body interactions and accounts for interactions at all frequencies of electromagnetic radiation. Interactions which involve, for instance, water on a substrate are seriously in error using the Hamaker microscopic approach where only electronic fluctuations are considered. A large contribution to the dielectric function for water arises from a Debye microwave relaxation term and infrared absorption terms in addition to the ultraviolet absorption term.[11]

Despite criticism of the term 'dispersion forces' and the recent macroscopic apporach to interfacial attraction, it is convenient for the purposes of this chapter to maintain this widely used nomenclature.

In considering the wetting of surfaces by coatings or adhesives, solid–liquid interactions rather than solid–solid interactions are those which must be considered. The types of interactions which are important in wetting are the short range ones which arise from intimate molecular contact. It has been shown[12] that if intimate molecular contact between two materials were possible, then van der Waals forces, even though they are the weakest of bonding forces, would provide an adhesive force which far exceeds the observed adhesion performance of the materials. The inability of liquids to wet completely a surface can be due to the lack of time allowed for the wetting process to take place if, for example, an adhesive is very viscous or the surface topography of a substrate allows intimate molecular contact of the adhesive only over a very small proportion of its surface. Between a liquid and areas of a surface which are not wetted there is an air gap, large on a molecular scale but probably no greater than ~ 10 nm, over which unretarded dispersion force interactions occur. However, the attractive forces between two materials in molecular contact (~ 0.5 nm) are greater than the attractive forces between the same materials separated by a gap of 10 nm by a factor of $\sim 10^4$; hence long range attractive forces play no part in interfacial adhesion. Therefore the strength of an interfacial bond depends on the ability of an adhesive to wet a solid substrate. Surface pretreatments are designed to change the thermodynamics or kinetics of wetting by liquids, so that, in the time allowed for the process, a larger proportion of the surface acquires molecular contact with the liquid. Surface pretreatments may also ensure that, where intimate molecular contact is achieved, the free energy of interaction is higher than that achieved by dispersion force interactions alone. Modification of the chemistry of polymer surfaces to incorporate functionalities which have a capability for hydrogen bonding is one way by which this can be achieved in practice.

3. RELATIONSHIP BETWEEN INTERMOLECULAR FORCES AND INTERFACIAL TENSION

Different types of intermolecular forces which operate in surfaces or at interfaces can be considered to be additive.[1] Hence, whereas the surface tension of alkanes arises only from dispersion forces acting at the surface, the surface tension of water is the result of contributions from polar forces and hydrogen bonding forces in addition to dispersion forces. However, Fowkes[13] suggested that since only like forces interact with like, the interaction between alkanes and more complex polar liquids and solids would be the result of dispersion force interactions alone, and that measurement of interfacial tension provided a means of determining the dispersion force contribution to the surface tension of that solid or liquid (γ^d).

The concept of an interfacial tension in molecular terms can be readily understood with reference to the description by Fowkes.[13] The interfacial region is generally confined to the thickness of one molecular layer. At the surface of any liquid which is in equilibrium with its vapour, this molecular layer will be in tension because the unopposed attraction of the bulk liquid causes an increase in the distance of intermolecular separation. Figure 1 shows the interface between two liquids which have been brought into contact. The surface tension of liquid 1 is γ_1, but when liquid 2 is brought into contact, the attraction of the surface layer of liquid 1 for liquid 1 is opposed by liquid 2; hence the tension in the surface layer of liquid 1 is reduced. A similar situation obtains for the

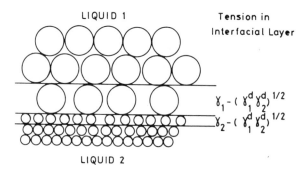

FIG. 1. Schematic arrangement of molecules at the interface between two liquids.

layer of liquid 2 which in the absence of liquid 1 has a surface tension γ_2. The decrease in tension of the monolayers caused by creating an interface between the two liquids is an average of the tensions and represents the interacting molecular forces. For dispersion force interactions across interfaces Fowkes used the geometric mean approximation of Berthelot. Consequently Fig. 1 illustrates that the tension in the interfacial layer of liquid 1 is $\gamma_1 - (\gamma_1^d \gamma_2^d)^{1/2}$. Similarly, the tension in the interfacial layer of liquid 2 is $\gamma_2 - (\gamma_1^d \gamma_2^d)^{1/2}$. The interfacial tension between liquids 1 and 2, γ_{12}, may be defined as the sum of the tensions in the outermost molecular layers:

$$\gamma_{12} = \gamma_1 + \gamma_2 - 2(\gamma_1^d \gamma_2^d)^{1/2} \qquad (1)$$

When considering solid/vapour or solid/liquid interfaces it is more usual to describe γ as a surface free energy rather than a surface tension. Since a liquid can interact with a solid in a similar way to the way it interacts with another liquid, the interfacial free energy between a liquid and a solid, γ_{SL}, can be written in the same form as eqn. (1) if the liquid and solid interact solely by dispersion forces:

$$\gamma_{SL} = \gamma_{SV^\circ} + \gamma_{LV^\circ} - 2(\gamma_S^d \gamma_L^d)^{1/2} \qquad (2)$$

where γ_{SV° refers to the surface free energy of the solid in equilibrium with the liquid vapour, and differs from the value in vacuo, γ_{S°, as shown in section 4.

The interfacial tension between two liquids was also derived from thermodynamic relations by Girifalco and Good.[14,15] Their expression differs from eqn. (1) in that an interaction parameter ϕ_{12} was included in the final term; ϕ_{12} was determined from molecular properties of the interacting liquids, such as the molecular polarisability, dipole moment and ionisation potential. For interacting liquids and solids where no specific interactions like hydrogen bonding are involved $\phi_{12} \leqq 1$. However, calculation of ϕ_{12} requires a knowledge of the exact chemical composition of the interacting surfaces, and is therefore difficult to apply to real adhesion situations which often involve interaction between multicomponent adhesives and chemically heterogeneous substrates.

4. CONTACT ANGLE AND THERMODYNAMIC EQUATIONS

The extent to which a liquid interacts with a solid to promote wetting is characterised by the contact angle (θ) which the liquid makes with the solid. If a drop of a liquid rests on a surface without completely wetting it, as shown in Fig. 2, the contact angle is the angle between the liquid/vapour interface and the liquid/solid interface from the point of three phase contact at equilibrium.

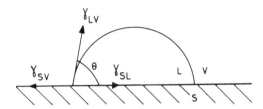

FIG. 2. Schematic diagram of a liquid drop on a surface, and forces acting at the line of three phase contact.

As long ago as 1805, Thomas Young[16] resolved the three surface tensions, represented by three coplanar vectors in Fig. 2, which act at any point of three phase contact on the surface to give

$$\gamma_{SV} - \gamma_{SL} = \gamma_{LV} \cos \theta \qquad (3)$$

Problems with experimental verification of Young's equation arise because there is no reliable way to measure γ_{SV} or γ_{SL} directly. In addition, equilibrium between solid, liquid and vapour is seldom reached because of internal stresses which frequently act in the outer surface layers of solids.

A more rigorous thermodynamic treatment of Young's equation was proposed by Bangham and Razouk,[17] who realised the importance of the adsorption of vapour of the liquid on the solid surface. The equi-

librium situation which the Young equation describes is that in which
the vapour is saturated

$$\gamma_{SV^\circ} - \gamma_{SL} = \gamma_{LV^\circ} \cos \theta \qquad (4)$$

where V° indicates the saturated vapour. Bangham and Razouk also
showed that, if the free energy of the solid in vacuo is γ_{S°, the equilibrium
spreading pressure of the saturated vapour (π_e) on the solid is given by

$$\pi_e = \gamma_{S^\circ} - \gamma_{SV^\circ} \qquad (5)$$

The contact angle of a liquid on a solid is controlled by the change in
free energy of adhesion of the solid and the liquid. Using a thermodynamic argument, Dupre[18] showed that the reversible work of separation
per unit area of solid and liquid (W_A) is equal to the change in free energy
per unit area of interface

$$W_A = \gamma_{S^\circ} + \gamma_{LV^\circ} - \gamma_{SL} \qquad (6)$$

Substitution of γ_{S° from equations (5) and (4) in equation (6) gives the
combined Young–Dupré equation

$$W_A = \gamma_{LV^\circ}(1 + \cos \theta) + \pi_e \qquad (7)$$

Experimental determinations of the spreading pressure of various liquids
on high energy surfaces such as metals and metal oxides from vapour
adsorption isotherms by Boyd and Livingston[19] and Harkins and co-
workers[20,21] showed that this contributed a large term to the work of
adhesion in eqn. (7). These data are conveniently tabulated by Zisman,[22]
who suggested that the spreading pressure of liquids on low energy
surfaces such as polymers provides a negligible contribution to the
reversible work of adhesion, provided that the contact angle of the liquid
on the surface is much greater than zero. Subsequent measurements on
low energy surfaces by Graham[23] and Whalen and co-workers[24,25]
supported this suggestion, although exceptions have been reported, for
instance, by Tadros and co-workers,[26] who measured a large spreading pressure for water on polyethylene. They suggested that water could
become structured at the interface and give rise to this effect. However,
Good's[27] interpretation of this result was that the polyethylene surface
was populated with some high energy sites which were sufficient to
generate a water film on the surface, but distributed in such a way that a
non-zero contact angle resulted. Good[27] concluded that in general, π_e is
small for a liquid that forms a non-zero contact angle on a solid,
provided that the surface is not appreciably rough or heterogeneous on a

microscopic scale. Obvious exceptions apply where molecules of the liquid chemisorb on the surface or where a low energy surface is populated by a number of high energy sites.

When an adhesive is applied as a liquid to a solid substrate, a thermodynamic work of adhesion may be calculated from eqn. (7), knowing the physical parameters of the system. However, this value is often several orders of magnitude lower than the strength of a joint comprising the substrate and solidified adhesive. In breaking the bond at a solid/solid interface, a large amount of work is required to produce permanent plastic deformation of the resulting fracture surfaces.[28] However, good wettability of surfaces is important for good adhesion; many correlations can be found in the literature between thermodynamic parameters of the adhesive/substrate system and the adhesive performance of joints. Some of these correlations will be examined in sections 6.1 and 7.1. It will be shown that use of the thermodynamic equations of Young and Dupré (eqns. (3), (6) and (7)) together with the relation derived by Fowkes (eqn. (2)), can provide estimates of γ_{SL} and the polarity of interacting surfaces. The effect of surface pretreatments which favourably change these parameters to enhance interfacial adhesion will also be discussed. In order to estimate thermodynamic parameters of surfaces, the careful measurement of contact angles of liquids on surfaces is a prerequisite.

5. CONTACT ANGLE MEASUREMENT

5.1. Methods of measurement

The contact angle used in Young's equation relates to the equilibrium angle of a liquid on a surface, because it is assumed that the surface is both smooth and chemically homogeneous. Hence equilibrium contact angles are those most commonly measured, where measurements are made from the profile of either a sessile drop of liquid (Fig. 3a), or an air bubble (Fig. 3b) resting on a planar surface. Despite the utmost care taken to prepare surfaces, they invariably contain a degree of roughness which is large compared with molecular dimensions, or some heterogeneity, so that an equilibrium contact angle in practice provides some average property of the surface. The contact angle may be obtained by drawing a tangent to the profile at the point of three phase contact where the drop profile has been enlarged either by image projection or photography. It may be measured directly using a telescope fitted with a

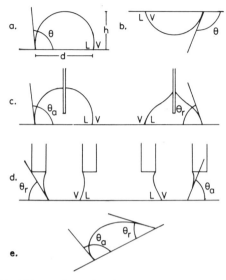

FIG. 3. Drop and bubble configuration for measurement of equilibrium, advancing and receding contact angles: (a) equilibrium sessile drop; (b) equilibrium pendant bubble; (c) advancing and receding drop; (d) advancing and receding captive bubble; (e) drop on tilted plate.

goniometer eyepiece, or indirectly by measuring the angle at which light from a point source is reflected from the surface of a liquid drop at its point of contact with the solid.[29] While the latter technique provides a rapid method of measurement, it is restricted to the measurement of contact angles less than 90°. Another useful method of indirect determination is measurement of the drop dimensions. Provided that the drops are sufficiently small for distortion of the drop due to gravity to be negligible, the contact angle may be calculated from the height h and base diameter d of the drop, which is considered to be the segment of a sphere

$$\tan \theta/2 = 2h/d \qquad (8)$$

This method is applicable to all drops which have a volume less than $10^{-10} \mathrm{m}^3$,[30] and can be extended to drops where $d < 0.5\,\mathrm{mm}$ and $\theta < 90°$.[31] The dependence of the profile of sessile drops on both contact angle and drop size is discussed by Padday[32] and reviewed by Johnson and Dettre.[33]

On real surfaces, the contact angle of a liquid may vary depending on whether the three phase line has been advanced or retracted over the surface. This defines advancing and receding contact angles where measurements are made immediately after the three phase line has come to rest. The most common methods of measurement are shown in Fig. 3(c)–(e). In Fig. 3(c), liquid is advanced or retracted over the surface by increasing or decreasing the volume which is dispensed by a needle from a syringe. The needle remains in the drop during measurements to avoid unnecessary vibration or distortion of the drop, and has no effect on the contact angle. The captive bubble technique is shown in Fig. 3(d), where the advancing and receding bubble provides the receding and advancing contact angle of the liquid respectively. In Fig. 3(e), the advancing and receding angles can be measured from a single drop at the limit when the drop begins to move on increasing the angle of tilt of the surface. Contact angles are measured as previously described for equilibrium angles. A number of other techniques can be used for measuring advancing and receding angles, e.g. wetting balance techniques, the capillary rise method and tilted plate method. These and other techniques are reviewed by Johnson and Dettre,[33] and more recently by Neumann and Good.[34] Contact angle measurements using different techniques are in good agreement, but it is more usual to use a sessile drop or captive bubble technique for measuring contact angles on planar substrates. One of the other techniques is more appropriate for surfaces which have complex geometries, such as fibres or capillaries.

5.2. Contact angle hysteresis

In the previous section it was suggested that the contact angle differed depending on whether the three phase line was advanced or retracted over the surface. The difference between the advancing angle and receding angle is known as the contact angle hysteresis. Zisman[35] has stated that, for thermodynamic purposes, it is essential to measure slowly advancing and slowly receding contact angles to ensure that the experimental conditions remain as close as possible to equilibrium. Under these conditions, Zisman found no contact angle hysteresis provided that the surface contained no pores or valleys into which the liquid could penetrate. However, besides apparent hysteresis which is caused by motion of the drop and represents a non-equilibrium situation, real hysteresis can exist on surfaces for a number of reasons. Good[3] has listed a number of causes for hysteresis which may occur under certain conditions: surface roughness, surface heterogeneity, diffusion, swelling, reorientation and fluid mechanical effects. The first two causes are the most important and have been studied in great detail, but the other causes are difficult to analyse because the kinetics of these processes are probably comparable with the speed of motion of the liquid across the surface.

5.2.1. Surface roughness

The roughness factor r of a surface is defined as the ratio of the true surface area, taking into account the troughs and ridges, to its apparent area, considering the surface as its planar projection. For a non-composite surface where a liquid can cover the entire surface without void formation, a drop of liquid will rest on the rough surface in a configuration which corresponds to a minimum in free energy. Under these conditions, the roughness ratio is given by Wenzel's equation,[36]

$$r = \frac{\cos \phi}{\cos \theta} \tag{9}$$

where ϕ is the apparent contact angle of the drop to the horizontal and θ is the intrinsic contact angle to the real surface. The thermodynamic derivation of eqn. (9) is given by Shuttleworth and Bailey,[37] while its limitations are discussed by Johnson and Dettre.[38] However, an examination of Wenzel's equation shows that, if the intrinsic contact angle on a given solid is less than 90°, then the apparent contact angle is reduced on a rough surface. On the other hand, an intrinsic contact angle greater than 90° causes the apparent contact angle to increase. Hence light abrasion of certain surfaces can enhance the wettability of surfaces and enhance adhesion to liquids (see Chapter 7).

The observed angle ϕ of a liquid on a rough surface is given by

$$\phi = \theta \pm \alpha \tag{10}$$

where α is the local slope of the surface at the point of contact of the liquid with the solid. The maximum and minimum observed angles occur when $\alpha = \alpha_{max}$, conditions which Shuttleworth and Bailey[37] associated with the advancing and receding contact angles. However, Johnson and Dettre[38] suggest that the presence of vibration in the system produces reproducible advancing and receding contact angles which are less than their extreme values, so that

$$\phi_a \lesssim \phi_{max} = \theta + \alpha_{max} \tag{11}$$

$$\phi_r \gtrsim \phi_{min} = \theta - \alpha_{max} \tag{12}$$

The advancing and receding contact angles on an idealised rough surface are represented schematically in Fig. 4(a). Liquids which have high intrinsic angles on solids may nót be able to penetrate very rough surfaces. Surfaces which are not completely penetrated are known as composite surfaces. These are only possible if the

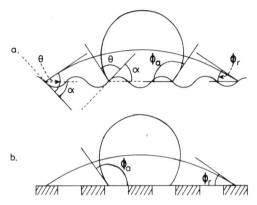

FIG. 4. Extreme drop configurations on idealised surfaces: (a) advancing and receding angles on non-composite rough surface; (b) advancing and receding angles on non-composite smooth patchwise-heterogeneous surface (hatched areas represent high energy sites, unhatched areas represent low energy sites).

slope of the ridges is such that

$$\theta = 180 - \alpha \tag{13}$$

On a composite surface where per unit apparent surface area the polymer–liquid interface has area Ω_{SL} and liquid–air interface under the drop has area Ω_{LV}, Cassie and Baxter[39,40] derived an equation for the observed contact angle where the free energy of the drop on the surface is minimised:

$$\cos \phi = \Omega_{SL} \cos \theta - \Omega_{LV} \tag{14}$$

Equation (14) reduces to Wenzel's equation (9) when $\Omega_{LV} = 0$, i.e. the liquid wets the entire surface. A most detailed study of rough surfaces was produced by Johnson and Dettre.[38] They proposed a theory for an idealised surface consisting of evenly spaced troughs and ridges with different roughness ratios, where the liquid front moved in a direction perpendicular to the ridges. They suggested that an idealised surface allowed drops of liquid to rest in a number of metastable configurations, each separated by an energy barrier. The height of this energy barrier is approximately proportional to the height of the asperity. They predicted that the energy barriers for composite surfaces are much lower than for non-composite surfaces. Experimenting on surfaces prepared with different degrees of roughness, Dettre and Johnson[41] found that these real systems behaved in a way which was predicted by these models. In particular they demonstrated the dramatic decrease in hysteresis which results when a non-composite surface becomes composite.

Although many pretreatments of polymer surfaces primarily involve chemical modification, in some cases this may also be accompanied by surface roughening, e.g. etching of polyolefin surfaces (see Chapter 9) and PTFE (see Chapter 10). However, pretreatments of metal surfaces are designed to modify the surface topography in one way or another; surface roughness in this context is considered in greater detail in Chapter 7.

5.2.2. Surface heterogeneity

On smooth surfaces, contact angle hysteresis is often still present, attributed to chemical heterogeneity. This occurs when different regions of the surface have different values of surface free energy. A surface can vary in its chemical composition; the surface of block copolymers consists of domains which have a surface free energy which is characteristic of the polymer comprising the different blocks. Even a homopolymer such as PTFE has a heterogeneous surface, the higher energy patches being provided by either carboxylic acid or short hydrocarbon chains resulting from initiator fragments which are able to survive the fabrication conditions. Even atoms or molecules which are chemically identical can produce different surface free energies depending on the arrangement of the atoms or molecules. Crystalline polymers and metals have a greater molecular or atomic density and hence a different surface free energy compared with their amorphous forms. Different crystallographic planes also exhibit different atomic packings, so that polycrystalline surfaces are also invariably heterogeneous.

For a flat surface consisiting of two types of domain 1 and 2 where the intrinsic contact angles are θ_1 and θ_2 respectively, Cassie[42] proposed that the equilibrium contact angle is the area-weighted average:

$$\cos \phi = \Omega_1 \cos \theta_1 + \Omega_2 \cos \theta_2 \tag{15}$$

Equation (15) is a general formulation of Cassie and Baxter's expression for the equilibrium angle on a composite rough surface given in eqn. (14);

when the second term is reduced to Ω_{LV} ($\cos \theta = 180°$) Johnson and Dettre[43] proposed a model for heterogeneous surfaces which is analogous to that for rough surfaces. They view advancing and receding angles on such surfaces as the result of a balance between vibration of the liquid and free energy barriers in the surface. They suggested that as the vibrational state of the liquid increases or the size of the heterogeneities decreases, the advancing and receding angles become closer to the equilibrium angle in eqn. (15). However, with smaller drop vibrations and larger heterogeneities, the advancing and receding angles approach the intrinsic angles θ_1 and θ_2. Hence the advancing angle provides a measure of the wettability of domains of lower surface free energy, whereas the receding angle relates to the wettability of domains of higher surface free energy. These situations are represented schematically in Fig. 4(b). Johnson and Dettre also suggested that the advancing and receding angles were independent of Ω_1/Ω_2 over a large range of values.

Johnson and Dettre reviewed experimental support for their theories, where studies of partial monolayer coverage of low energy waxes on high energy metallic surfaces were cited. The theory qualitatively fits observations, and suggests that, on real surfaces, depleted monolayers adsorb in a patchwise manner. Studies include advancing contact angle measurements of hexadecane on a surface of chromium or stainless steel with an adsorbed layer of perfluoro-octanoic acid[44] and methylene iodide contact angles on depleted monolayers of stearic acid on chromium, platinum and mica.[45]

Good and Koo[46] have invoked the heterogeneous nature of surfaces to explain the contact angle dependence on the size of sessile drops used to make the measurement. They found that for water on a tetrafluoroethylene/hexafluoropropylene copolymer (TFE/HFP) and on poly(methyl methacrylate) both the advancing and receding angles decreased rapidly as the drop diameter decreased below 4 mm. However contact angles of decane on TFE/HFP were independent of drop size. They proposed that if the polymer surface was heterogeneous so that water wetted the high energy sites, the perimeter of a water drop sitting on the surface would be tortuous. When the drop is small, the tortuosity of the three phase contact line is sufficient to reduce the macroscopic contact angle and to account for the observed drop size dependence on contact angle. Decane wets neither type of surface site; it therefore provides a non-tortuous three phase contact line. The results cannot be explained on the basis of surface roughness.

Oriented polymer surfaces have been studied by Good *et al.*[47] to confirm that this type of heterogeneity resulted in contact angle hysteresis. They found that oriented films of PTFE, TFE/HFP copolymer, PP and PE displayed anisotropic contact angles, where measurements as a liquid front moved perpendicular to the direction of stretch gave higher results than the same measurements parallel to the direction of stretch. They showed that surface roughness could not account for the observed contact angle hysteresis or anisotropy.

5.3. Sensitivity of contact angle measurements to surface composition and structure

Contact angle measurements probe the first molecular layer of surface and hence have a sensitivity to molecular organisation in the surface which is far greater than surface analytical techniques such as MIR or XPS. Contact angle measurements can therefore provide a wealth of information about surfaces if data are correctly interpreted and used in combination with other techniques.

The previous section showed how microscopic changes on a surface, for instance by changing the asperity height on a homogeneous surface or the introduction of a small number of high energy sites on a low energy surface, can dramatically change a macroscopic measurement such as contact angle.

There are a number of other phenomena which can also be monitored by measurements of contact angles. Schrader[48] has shown that a finite water contact angle on high energy surfaces such as glasses, ceramics, metals and oxides represents a sensitive method for detecting surface contamination. Clean high energy surfaces rapidly acquire adsorbed organic films after exposure to air, considerably lowering the surface free energy. Using the Lifshitz theory, Parsegian *et al.*[49] calculated the free energy due to long range interactions between water and a gold surface contaminated by a variable thickness of tetradecane (model for the contaminant) and compared this with the short range interaction of a water film on a contaminated gold surface. They showed that the long range contribution became comparable with short range interactions for tetradecane thicknesses <2Å, or considerably less than a monolayer thickness. These calculations demonstrated that long range forces due to metallic surfaces do not significantly extend past one monolayer of an organic contaminant and could not allow water to spread on a contaminated surface.

Some very elegant measurements by Neumann and co-workers[50,51] demonstrated the precision with which contact angle measurements could be made. This involved measuring the capillary rise of a liquid at a vertical plate where the height could be measured by a cathetometer with a precision of 2×10^{-6}m. Measurements of capillary rise as a function of temperature produced sharp changes in slope of the curve at temperatures which corresponded to various crystalline transitions. Measurements on PET[51] showed transitions at 75°C which corresponded to T_g and also at 30°C. The latter transition had not been previously reported in the literature but was subsequently confirmed by DSC measurements.

Contact angle measurements can also be used to monitor the reorientation of polymer molecules at surfaces. If polymer chains have a sufficiently high mobility, they will reorientate so that they present to air a surface which corresponds to a minimum in surface free energy. Baszkin and Ter-Minassian-Saraga[52] showed that contact angles on chemically oxidised polyethylene increased after heating the polymer to 80°C. They suggested that this was consistent with reorientation of the PE molecules in such a way that the carboxylic acid groups were directed into the bulk of the film. A similar conclusion was reached by Briggs et al.[53] in the case of corona discharge treated PET. The contact angle of water increased with increasing time of ageing of the surface.

Spectroscopic measurements by XPS, peel strength measurements of an autohesive seal, and contact angle measurements, together provided evidence for the migration of low molecular weight oxidised material away from the surface, accompanied by reorientation of higher molecular weight polymer chains in the surface at room temperature.

6. WETTING OF SURFACES

6.1. Polymers
When a liquid spreads on a surface, a solid–liquid interface and liquid–vapour interface are formed at the expense of a solid–vapour interface. The free energy change for formation of the solid–liquid interface is defined as the spreading coefficient

$$S = \gamma_{SV^\circ} - \gamma_{SL} - \gamma_{LV^\circ} \qquad (16)$$

For polymers, $\gamma_{SV^\circ} < 60 \, \text{mJ m}^{-2}$ so that liquids will spread on polymers ($S > 0$) only if γ_{LV° is small. Fox and Zisman[54] measured the *equilibrium* contact angle of a homologous series of pure liquids on smooth low energy surfaces and found a linear relationship between cos θ and the surface tension of the liquid γ_{LV°. They defined the 'critical surface tension of wetting' of a solid γ_c as the value of γ_{LV° at which a liquid just wets the surface with zero contact angle. This can be obtained graphically from the intercept of the plot of cos θ against γ_{LV° with the line cos $\theta = 1$.

A review of the pioneering work of Zisman and his co-workers[35] shows how thoroughly the wettability of different low energy surfaces has been studied. Table 1 provides values of γ_c for some common polymeric surfaces. This table shows that the values of γ_c for various polymers can be correlated qualitatively with the molecular structure of their mono-mers. Successive replacement of hydrogen atoms by fluorine atoms in ethylene decreases the critical surface tension of wetting of the polymer. An increase in γ_c relative to polyethylene can be achieved by replacement of hydrogen atoms by more polar atoms or groups or by the insertion of polar molecules into the polymer chain. Contact angle measurements used to determine γ_c or to estimate thermodynamic parameters of surfaces or interfaces (see section 7) should be carried out on smooth surfaces using pure liquids which are allowed to reach equilibrium with the vapour phase. There is no justification for using liquid mixtures for these measurements.[3]

On certain surfaces, Ellison and Zisman[55] found different values of γ_c for the same solid, depending on whether they used a series of non-polar, polar non-hydrogen bonding or hydrogen bonding liquids. This prompt-ed Kitazaki and Hata[56] to investigate these observations in more detail. They measured contact angles of different series of liquids on partially fluorinated polymers and predicted which value of γ_c was closest to the intrinsic surface free energy γ_{S°. From Young's equation (eqn. 4), the relationship between γ_c and γ_{S° can be obtained by putting $\gamma_{LV^\circ} = \gamma_c$ and $\gamma_{SL} = \gamma_{SL}^*$ at cos $\theta = 1$ and neglecting π_e

$$\gamma_c = \gamma_{S^\circ} - \gamma_{SL}^* \qquad (17)$$

where γ_{SL}^* depends on the combination of liquid and solid. Generally the value of γ_{SL}^* is a minimum when the solid and liquid have the same polarity. For polytrifluoroethylene, liquids which are capable of hydro-gen bonding give values of γ_c which are closest to γ_{S°; the highly acidic protons in the polymer provide sites for hydrogen bonding.

Although thermodynamic parameters such as γ_{S° are not accessible to direct experimental measurement, γ_c can be readily measured. If the correct series of liquids is chosen such that γ_c is close to γ_{S°, then γ_c is a useful parameter for correlating with adhesive strength. Kitazaki and

TABLE 1
CRITICAL SURFACE TENSION OF WETTING, AND SURFACE FREE ENERGY FOR POLYMERIC SOLIDS

Polymer	Chemical structure of monomer compared with ethylene	γ_c at 20°C (mN m^{-1})[22]	γ_s^d (mJ m^{-2})	γ_s^p (mJ m^{-2})[69]	γ_s^o (mJ m^{-2})
Polytetrafluoroethylene	4H replaced by F	18·5	18·6	0·5	19·1
Polytrifluoroethylene	3H replaced by F	22	19·9	4·0	23·9
Poly(vinylidene fluoride)	2H replaced by F	25	23·2	7·1	30·3
Poly(vinyl fluoride)	1H replaced by F	28	31·3	5·4	36·7
Low density polyethylene	—	31	33·2	—	33·2
Polypropylene[a]	1H replaced by CH$_3$	31	30·2	—	30·2
Poly(methyl methacrylate)	1H replaced by ester	39	35·9	4·3	40·2
Poly(vinyl chloride)	1H replaced by Cl	39	40·0	1·5	41·5
Poly(vinylidene chloride)	2H replaced by Cl	40	42·0	3·0	45·0
Polystyrene	1H replaced by benzene ring	43	41·4	0·6	42·0
Poly(ethylene terephthalate)	Polar monomers inserted	43	43·2	4·1	47·3
Poly(hexamethylene adipamide)	(ester, amide) in hydrocarbon chain	46	35·9	4·3	40·2

[a] Data for polypropylene from ref. 71.

Hata[56] took γ_c values of a number of polymers which were obtained from a series of hydrogen bonding liquids and plotted the tensile shear strength of epoxy adhesives against γ_c for different polymers using the data of Schonhorn et al.[57,58] and Levine et al.[59] The adhesive strength was found to be a maximum for polymers having values of γ_c closest to the surface tension of the adhesive, giving support to the widely held criterion for optimum adhesion that $\gamma_{LV^\circ} = \gamma_c$.

The value of γ_{LV° for aqueous coatings can be lowered by the addition of surface active agents to a value below the critical surface tension of wetting of all polymer surfaces. Although surface active agents containing a hydrocarbon chain as the hydrophobic part of the molecule cannot lower the surface tension of water below $\sim 25\,\mathrm{mN\,m^{-1}}$, perfluorinated surface active agents can lower the surface tension of water to $15\,\mathrm{mN\,m^{-1}};^{60}$ this is adequate for spreading on fluorinated surfaces. Solvent-borne coatings generally have low values of γ_{LV° and spread readily on low energy surfaces. Some problems are encountered when applying polar adhesives such as epoxides to polyolefin substrates. In the liquid state, epoxides have a surface tension in the region $35-45\,\mathrm{mN\,m^{-1}}$. Consequently they will not spread on polyethylene, which has a critical surface tension for wetting of $31\,\mathrm{mN\,m^{-1}}$. Spreading will occur only if the surface is pretreated to raise the surface free energy above that of the adhesive. However, Sharpe and Schonhorn[61] showed that if polyethylene were melted into a solid epoxide surface ($\gamma_c = 32 \cdot 9\,\mathrm{mN\,m^{-1}}$) Zisman's criterion for wetting was satisfied, good wetting occurred, and a high bond strength resulted.

6.2. Metals

Metal surfaces are regarded as high energy surfaces, in that their surface free energy γ_{S° is in the region of $1\,\mathrm{J\,m^{-2}}$. However, all high energy surfaces readily adsorb contaminant organic molecules or water vapour from the air, so that by reference to eqn. (16) γ_{SV° can be reduced to such a low value that $S < 0$. Under these conditions, liquids will show finite contact angles on the surface. A review[62] of the controversy over whether a metal which was resistant to surface oxidation such as gold was completely wetted or not by water, demonstrates the degree of cleanliness of metal surfaces required in order to obtain reproducible results. However, early work by Bewig and Zisman,[63] which demonstrated the complete wetting of pure gold and platinum by water under conditions which ensured the elimination of all traces of organic impurities from the atmosphere, was confirmed by Schrader.[64] He measured the contact angle of water on metals after surface preparation in ultra-high vacuum. Using the same techniques Schrader[65] also found zero contact angles for water on silver and copper films.

Zisman[62] measured the critical surface tension for wetting of a number of different metal and metal oxide surfaces which were exposed to atmospheres with controlled humidity. It was found that, irrespective of the nature of the substrate, γ_c for all surfaces was lowered to $\sim 45 \, \text{mN m}^{-1}$ at 0·6%RH and to $\sim 37 \, \text{mN m}^{-1}$ at 95%RH. Hence for any clean, smooth hydrophilic surface such as a metal or its oxide, exposure to a humid atmosphere converts a high energy surface to a low energy surface, the surface energy being dependent on the concentration of adsorbed water molecules. The surface free energies of metals and their oxides are sufficiently high to allow good wetting by any adhesive, and one of the purposes of a pretreatment is to remove surface contamination (see Chapter 7 and 8).

The strong affinity of organic molecules for high energy surfaces is important for mould-release applications.[27] A monolayer of a long-chain aliphatic acid such as stearic acid or a film of polydimethylsiloxane can be deposited on to a metal, lowering the critical surface tension to $24 \, \text{mN m}^{-1}$ at 20°C. The coatings convert the high energy surface into a low energy surface. Such abhesive coatings are more effective the lower their surface free energy. A liquid which is poured into a mould coated with an abhesive layer produces such a high contact angle that poor adhesion results, and on solidification, the mould and moulding may be easily parted by application of a small external stress.

6.3. Thermodyanamics versus kinetics: wetting criteria for good adhesion

The necessity for intimate molecular contact between a liquid and a surface in order to achieve good adhesion is interpreted by Zisman,[22] in terms of obtaining complete wetting as indicated by zero contact angle. However, this criterion assumes that the surface is ideal. Surfaces of practical interest are seldom completely smooth; this observation led Huntsberger[66,67] to re-examine the thermodynamics of wetting. Huntsberger derived a thermodyamic equation expressing the free energy change in going from a non-wetted to a completely wetted state:

$$\Delta G = -\gamma_{LV} \left(1 + (\Omega_{SV}/\Omega_{LV}) \cos \theta\right) \qquad (18)$$

where Ω_{SV}/Ω_{LV} is the ratio of the actual to the apparent surface area. From eqn. (18) the wetted state is thermodynamically favoured, as given by a negative free energy change, unless $(\Omega_{SV}/\Omega_{LV}) \cos \theta < -1$. This suggests that any smooth surface at equilibrium is completely wetted by

D. G. RANCE

all liquids. This does not mean that the liquids will spread on the surface because, for many combinations of solid and liquid, the equilibrium contact angle will be high. Since $\cos \theta$ is positive for $\theta < 90°$, it also follows from eqn. (18) that any rough surface will be completely wetted at equilibrium provided that the intrinsic contact angle of the liquid is $< 90°$. These considerations led Huntsberger to propose that poor interfacial adhesion resulted not from thermodynamic but kinetic aspects of adhesives application. Huntsberger suggested that at first sight, the rate of wetting of a surface should be proportional to $\cos \theta$, giving a maximum rate of wetting with liquids for which $\theta = 0$. However, it was also demonstrated that the maximum rate of wetting could occur with liquids showing non-zero contact angles; this emphasised the important role of both surface and interfacial free energy in determining the rate of wetting. Achieving a maximum rate of wetting is important in practice because the high viscosity of many adhesives and the short time they remain in the liquid state after contact with a substrate limit their penetration into troughs on the surface.

7. ANALYSIS OF INTERFACIAL INTERACTIONS FROM CONTACT ANGLE MEASUREMENTS

7.1. Surface and interfacial free energy

Polymer surfaces, apart from untreated polyolefins, can interact with liquids by attractive forces other than dispersion forces. Even polyolefin surfaces can participate in stronger interactions if pretreated in such a way that the surface chemistry is changed. Most pretreatments for polyolefins cause oxidation, introducing polar groups into the surface (see Chapter 9). Fowkes[68] suggested that the total surface free energy of a solid or liquid is the sum of different intermolecular forces, for example dispersion, polar, hydrogen bonding

$$\gamma = \gamma^d + \gamma^p + \gamma^h + \ldots \tag{19}$$

Where both interacting materials exhibit intermolecular forces of a given type, the interaction will contribute a term to the work of adhesion;

$$W_A = W_A^d + W_A^p + W_A^h + \ldots \tag{20}$$

For cases where both a solid and a liquid are polar, Owens and Wendt[69] and Kaelble and Uy[70] suggested that all polar interactions, including more specific hydrogen bonding interactions, could be combined to provide one term which simplifies eqn. (19) to the sum of two terms, γ^d and γ^p. By assuming that interactions between polar forces could be approximated by a geometric mean approximation in the same way as dispersion forces, they obtained an expression for the interfacial tension between a liquid and solid by an extension of the Fowkes equation (2) :

$$\gamma_{SL} = \gamma_{SV^\circ} + \gamma_{LV^\circ} - 2(\gamma_S^d \gamma_L^d)^{1/2} - 2(\gamma_S^p \gamma_L^p)^{1/2} \tag{21}$$

By combining eqn. (21) with the Young and Dupre equations (4) and (6) and by considering the surface pressure to be negligible,

$$\gamma_{LV^\circ}(1 + \cos\theta) = 2(\gamma_S^d \gamma_L^d)^{1/2} + 2(\gamma_S^p \gamma_L^p)^{1/2} \tag{22}$$

The publication by Fowkes[68] of values of γ_L^d for a series of liquids from measurements of interfacial tension with alkanes which interact with other liquids by dispersion force attraction, led Owens et al.[69,71] and Kaelble[70] to take $\gamma_L^p = \gamma_{LV^\circ} - \gamma_L^d$. The measurement of equilibrium contact angles of two of these liquids on a given solid provided all the information needed to solve two simultaneous equations constructed using eqn. (22) for γ_S^d and γ_S^p. The sum of these components provided an estimate of the free energy of the solid, provided that the spreading pressure was indeed negligible.

Values of $\gamma_S^p \gamma_S^d$ and γ_{S° obtained in this way from contact angles of water and methylene iodide on a number of common polymers are shown in Table 1. A comparison of values of γ_{S° with Zisman's values of γ_c shows that for a given material the values are close and in most cases $\gamma_{S^\circ} > \gamma_c$ as predicted by eqn. (17). Values of γ_{LV° and γ_L^d for a number of useful test liquids are given in Table 2. Values of γ_S^p and γ_S^d for solid surfaces can also be determined from contact angles of a number of well-characterised liquids by a graphical method.[74]

The resolution of the surface free energy into two components provides a qualitative method for following changes in surface chemical composition. Krüger and Potente[74] have measured values of γ_S^d and γ_S^p of polypropylene films as a function of the degree of corona discharge treatment. Figure 5 shows that γ_S^p increases with increase in energy in the discharge. The polar groups which are introduced by corona discharge treatment are important for making good adhesive joints with a polyurethane adhesive as indicated by the peel strength measurements in Fig. 5.

Figure 6 shows the change in γ_S^p and γ_S^d obtained from measurement of water and formamide contact angles on a PET surface which was aged after corona discharge treatment with an energy dissipation of

144   D. G. RANCE

TABLE 2
SURFACE FREE ENERGY OF VARIOUS LIQUIDS AT 20°C

Liquid	γ_{LV° $(mJ\,m^{-2})$	γ_L^d $(mJ\,m^{-2})$	$\gamma_{LV^\circ}-\gamma_L^d$ $(mJ\,m^{-2})$	Reference
Water	72·8	21·8	51·0	68
Glycerol	63·4	37·0	26·4	68
Formamide	58·2	39·5	18·7	68
Methylene iodide	50·8	49·5	1·3	69
	50·76	(50·76)	—	72
Ethane-1,2-diol	48·3	29·3	19·0	73
Dimethyl sulphoxide	43·54	34·86	8·68	72
Tricresyl phosphate	40·70	36·24	4·46	72
Pyridine	38·00	37·16	0·84	72
Dimethyl formamide	37·30	32·42	4·88	72
2-Ethoxyethanol	28·6	23·6	5·0	73
n-Hexadecane	27·6	(27·6)	—	68
Dimethylsiloxanes	19·0	16·9	2·1	68

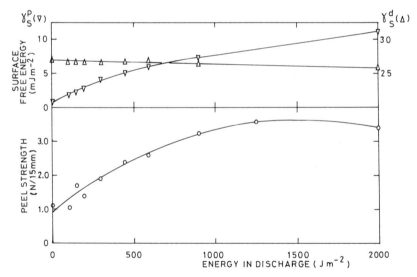

FIG. 5. Polar and dispersion force contribution to surface free energy of a discharge treated polypropylene copolymer surface and peel strength with a polyurethane resin against energy dissipated in the discharge. Reproduced from Kruger and Potente, *J. Adhesion*, **11** (1980), 113–124 by permission of the publishers, Gordon and Breach Science Publishers Inc. ©.

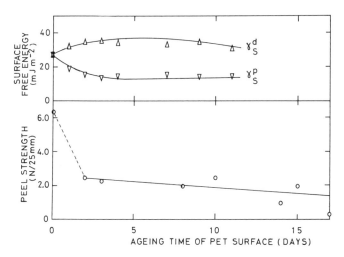

FIG. 6. Polar and dispersion force contribution to surface free energy of a discharge treated poly(ethylene terephthalate) surface (energy in discharge 6200 Jm^{-2}) and peel strength of an autohesive seal against ageing time. Reproduced from Briggs, Rance, Kendall and Blythe, *Polymer* **21** (1980), 895–900 by permission of the publishers, IPC Business Press Limited ©.

$6200\,Jm^{-2}$. The ageing of the PET surface was attributed[53] to the migration of low molecular weight oxidised species from the surface together with reorientation of molecules with higher molecular weight to produce intramolecular hydrogen bonds. These processes result in a decrease in the surface concentration of polar groups. This is reflected by a decrease in γ_S^p and a decrease in the peel strength of an autohesive seal.

Calculation of the components of the surface free energy of solids from contact angle measurements led Wu[75] and Potente and Krüger[76] to analyse solid/solid interactions in an adhesive joint. For two solids B and C, the interfacial free energy γ_{BC} from eqn. (21) becomes

$$\gamma_{BC} = \gamma_B + \gamma_C - 2(\gamma_B^d\gamma_C^d)^{1/2} - 2(\gamma_B^p\gamma_C^d)^{1/2} \qquad (23)$$

Equation (23) is just one of a number of different expressions for the interfacial free energy between two materials.[75] However, the values obtained from these analyses have all been found to be less than those experimentally measured for polymer melts.[77]

Interfacial free energy between an adhesive and its substrate has been identified[78] as perhaps the most important parameter governing interfacial adhesion. In general, practical adhesive strength is a maximum only when the interfacial free energy is a minimum. In section 6.1 it was suggested that matching the surface free energy of the substrate and adhesive would produce good interfacial adhesion. Expressing this in terms of eqn. (23), $\gamma_B = \gamma_C$. Simultaneous minimisation of the interfacial free energy (i.e. $\gamma_{BC} = 0$) can only occur if $\gamma_B^d = \gamma_C^d$, which implies $\gamma_B^p = \gamma_C^p$. This suggests that good interfacial adhesion will be achieved for an adhesive and substrate which both have the same surface free energy if their polarity is matched. Potente and Krüger[76] have shown that this criterion is valid for acrylic and alkyd resin paints which have been applied to a variety of polymer surfaces.

The literature contains many other criteria besides minimising interfacial tension for optimising interfacial adhesion. These include maximising the theoretical work of adhesion, optimising the spreading coefficient, matching solubility parameters, etc.; they have all been extensively reviewed by Mittal.[79]

7.2. Acid–base interactions

Expressions for the work of adhesion between polar materials considered so far have contained only two terms, one W_A^d arising from dispersion forces, the other W_A^p from polar interactions. This treatment allowed calculation of thermodynamic parameters such as surface free energy and its components, and interfacial free energy as outlined in the previous section. However, Fowkes[80] pointed out that all three terms which contribute to the work of adhesion in eqn. (20) must be considered, since the contribution of W_A^h to the total work of adhesion could be large where a surface and a liquid have the ability to form hydrogen bonds. Polymer surfaces can incorporate chemical groups which can have one of three types of hydrogen bonding capability:

(a) proton acceptors (electron donor or basic) such as esters, ketones, ethers or aromatics;

(b) proton donor (electron acceptor or acidic) such as partially halogenated molecules;

(c) both proton acceptor and proton donor such as amides, alcohols and amines.

Fowkes[80] analysed data published by Dann[73] on the contact angle of different liquids of type (c) on a number of polymer surfaces. A plot of $W_A - W_A^d$ for these liquids on polyester and polyamide surfaces against $\gamma_{LV°} - \gamma_L^d$ produced a linear relationship. This implies that the 'polar' contribution to the work of adhesion is proportional to the first power of the 'polar' contribution to the surface free energy, not to power $0·5$ as given by the mean square relationship suggested by Kaelble and Uy, and by Owens and Wendt, to describe these interactions.

Fowkes[81] has properly considered specific interactions such as hydrogen bonding at interfaces, building upon the work of Drago and co-workers,[82,83] who regarded hydrogen bonding as a type of acid–base interaction. Drago suggested that intermolecular interactions in solution could be approximated by the sum of dispersion force interactions and acid–base interactions; contributions due to dipole–dipole interactions were considered negligible. Drago measured the enthalpy of acid–base interactions ΔH^{AB} for a number of Lewis acids and Lewis bases in a neutral solvent such as carbon tetrachloride, characterising each acid A and base B by two constants C and E such that

$$-\Delta H^{AB} = C_A C_B + E_A E_B \qquad (24)$$

Fowkes[84] expressed the work of adhesion between two materials where specific interactions are important as

$$W_A = W_A^d + W_A^{AB} + W_A^p \qquad (25)$$

$$W_A = 2(\gamma_A^d \gamma_B^d)^{1/2} - f(C_A C_B + E_A E_B)n^{AB} + W_A^p \qquad (26)$$

where n^{AB} is the number of moles of acid/base pairs per unit area and $f \simeq 1$ converts enthalpy/unit area into surface free energy.

Fowkes and Maruchi[72] have measured contact angles of a number of acidic liquids, such as mixtures of phenol in tricresyl phosphate, and basic liquids, such as pyridine, dimethylformamide and dimethyl sulphoxide on acidic ethylene/acrylic acid copolymer surfaces and basic ethylene/vinyl acetate copolymer surfaces. These experiments were designed to test Drago's assumption that the contribution of W_A^p in eqn. (25) is negligible and to assess the strength of acid–base interactions in terms of the contribution of W_A^{AB} to the total work of adhesion. Combining eqns. (7) and (25) for various solid–liquid combinations Fowkes calculated

$$W_A - W_A^d - \pi_e = \gamma_{LV^\circ}(1 + \cos\theta) - 2(\gamma_s^d \gamma_L^d)^{1/2} \qquad (27)$$

Values of γ_S^d for each polymer were calculated from the contact angle of methylene iodide on polymer surfaces and γ_L^d from the contact angle of each test liquid on the surface of polyethylene (some γ_L^d values from this work appear in Table 2). Methylene iodide and polyethylene were considered as standard materials which interact with other materials by dispersion force attraction. Figure 7 shows values of $W_A - W_A^d - \pi_e$ for acidic liquids and a basic liquid on basic surfaces. The data for pyridine show that there is no contribution to the work of adhesion other than that due to dispersion force attraction, despite the highly polar nature of the interacting materials. This supports the suggestion that the completely dipolar contribution to the work of adhesion, W_A^p, is indeed negligible. Hence the data for acidic liquids on ethylene/vinyl acetate copolymer surfaces in Fig. 7 indicate the magnitude of acid–base interactions, which increase both with increasing phenol content in the tricresyl phosphate and increasing vinyl acetate content in the copolymer. Fowkes and Maruchi also reported consistent findings for acidic surfaces.

Equation (26) provides a quantitative value for the number of acid–base pairs which interact. On this basis Fowkes suggested that most of the acidic or basic groups on the polymers studied remained buried in

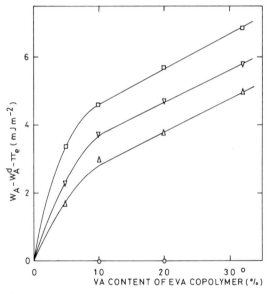

FIG. 7 $W_A - W_A^d - \pi_e$ from contact angle measurements of acidic and basic liquids on basic ethylene/vinyl acetate copolymer surfaces against copolymer composition ◯, Pyridine; △, tricresyl phosphate + 35% phenol: ▽, tricresyl phosphate + 48% phenol; ▢, tricresyl phosphate + 72% phenol. Reproduced from Fowkes in *Polymer science and technology*, Vol. 12A, Adhesion and adsorption of polymers, L.-H. Lee (Ed.), 1980, pp. 43–52 by permission of the publishers. Plenum Press Limited ©.

the surface. This is consistent with the surface having been able at some time to minimise its surface free energy by molecular reorientation.

A number of studies by Fowkes and co-workers[81] suggest the applicability of the acid–base concept of interfacial interactions to a wide range of industrial problems; tailoring the chemistry of surfaces to enhance ink and lacquer adhesion, enhancing the adhesion between filler particles and polymer matrices, increasing the adhesion of polymer coatings to metal oxides, improving the water resistance of coated metals. While Drago and co-workers have provided data for many acids and bases, there are still gaps in their tables which need to be filled in order to predict confidently the acid–base interactions between a wider range of materials. However, this approach provides a very useful way for estimating the interfacial interaction energy between materials because it includes in its analysis specific hydrogen bonded interaction. In conclusion, the acid–base concept of interfacial adhesion appears to be the most useful for predicting the practical adhesion between surfaces where hydrogen bonding plays an important role. It is therefore particularly applicable in the context of this book since so many methods of surface pretreatment for plastics produce an increase in hydrogen bonding capability.

ACKNOWLEDGEMENTS

The author wishes to thank Dr D. Briggs and Mr E. L. Zichy for reading the manuscript and for their many helpful comments.

REFERENCES

1. Fowkes, F. M., *J. Phys. Chem.*, **66** (1962), 382.
2. Fowkes, F. M., *J. Phys. Chem.*, **67** (1963), 2538.
3. Good, R. J. in *Surface and colloid science*, Vol. 11, R. J. Good and R. R. Stromberg (Eds.), Plenum, New York–London, 1979, p. 1.
4. Good, R. J., *J. Coll. Interface Sci.*, **59** (1977). 398.
5. Hamaker, H. C., *Physica*, **4** (1937), 1058.
6. Gregory, J., *Adv. Coll. Interface Sci.*, **2** (1969), 396.
7. Visser, J., *Adv. Coll. Interface Sci.*, **3** (1972), 331.
8. Tabor, D. and Winterton, R. H. S., *Nature*, **219** (1968), 1120.
9. Dzyaloshinskii, I. E., Lifshitz, E. M., and Pitaevskii, L. P., *Adv. Phys.*, **10** (1961), 165.
10. Lifshitz, E. M., *Soviet Phys. JETP*, **2** (1956), 73.
11. Gingell, D., and Parsegian, V. A., *J. Coll. Interface Sci.*, **44** (1973), 456.

12. Tabor, D., *Rep. Progr. Appl. Chem.*, **36** (1951), 621.
13. Fowkes, F. M. in *Contact angle wettability and adhesion, ACS. Adv. Chem. Ser.*, **43** (1964), p. 99.
14. Girifalco, L. A. and Good, R. J., *J. Phys. Chem.*, **61** (1957), 904.
15. Good, R. J. and Girifalco, L. A., *J. Phys. Chem.*, **64** (1960), 561.
16. Young, T., *Phil. Trans. Roy. Soc. (London)*, **95** (1805), 65.
17. Bangham, D. H. and Razouk, R. I., *Trans. Faraday Soc.*, **33** (1937), 1459.
18. Dupré, A., *Theorie mechanique de la chaleur*, Gauthier-Villars, Paris 1869, p. 369.
19. Boyd, G. E. and Livingston, H. K., *J. Amer. Chem. Soc.*, **64** (1942), 2383.
20. Harkins, W. D. and Loeser, E. H., *J. Chem. Phys.*, **18** (1950). 556.
21. Loeser, E. H., Harkins, W. D. and Twiss, S. B., *J. Phys. Chem.*, **57** (1953), 251.
22. Zisman, W. A. in *Contact angle, wettability and adhesion, ACS Adv. Chem. Ser.*, **43** (1964), 1.
23. Graham, D. P., *J. Phys. Chem.*, **68** (1964), 2788; **69** (1965), 4387.
24. Whalen, J. W., *J. Coll. Interface Sci.*, **28** (1968), 443.
25. Wade, W. H. and Whalen, J. W., *J. Phys. Chem.*, **72** (1968), 2898.
26. Tadros, M. E., Hu, P. and Adamson, A. W., *J. Coll. Interface Sci.*, **49** (1974), 184.
27. Good, R. J., *J. Coll. Interface Sci.*, **52** (1975), 308.
28. Cherry, B. W., Aspects of Surface Chemistry and Morphology, in *Plastics, surface and finish*, S. H. Pinner and W. G. Simpson (Eds), Butterworth, London, 1971, p. 217–42.
29. Fort, T. and Patterson, H. T., *J. Coll. Sci.*, **18** (1963), 217.
30. Mack, G. L., *J. Phys. Chem.*, **40** (1936), 159.
31. Bartell, F. E. and Zuidema, H. H., *J. Amer. Chem. Soc*, **58** (1936), 1449.
32. Padday, J. F., *Proc. Roy. Soc. Lond. A*, **330** (1972), 561.
33. Johnson, R. E. and Dettre, R. H., in *Surface and colloid science*, Vol. 2, E. Matijevic (Ed.), John Wiley and Sons Inc., New York, 1969, p. 85.
34. Neumann, A. W. and Good, R. J., *Surf. Coll. Sci.*, **11** (1979), 31.
35. Zisman, W. A., *J. Paint Technol.*, **44** (1972), 42.
36. Wenzel, R. N., *Ind. Eng. Chem.*, **28** (1936), 988; *J. Phys. Chem.*, **53** (1949), 1466.
37. Shuttleworth, R. and Bailey, G. L. I., *Disc. Faraday Soc.*, **3** (1948), 16.
38. Johnson, R. E. and Dettre, R. H. in *Contact angle wettability and adhesion, ACS Adv. Chem. Ser.*, **43** (1964), 112.
39. Cassie, A. B. D. and Baxter, S., *Trans. Faraday Soc.*, **40** (1944), 546.
40. Baxter, S. and Cassie, A. B. D., *J. Text. Inst.*, **36** (1945), T67.
41. Dettre, R. H. and Johnson R. E. in *Contact angle wettability and adhesion, ACS Adv. Chem. Ser.*, **43** (1964), 136.
42. Cassie, A. B. D., *Disc. Faraday Soc.*, **3** (1948), 11.
43. Johnson, R. E. and Dettre, R. H., *J. Phys. Chem.*, **68** (1964), 1744.
44. Shepard, J. W. and Ryan, J. P., *J. Phys. Chem.*, **63** (1959), 1729.
45. Gaines, G. L. Jr., *J. Coll. Sci.*, **15** (1960), 321.
46. Good, R. J. and Koo, M. N., *J. Coll. Interface Sci.*, **71** (1979), 283.
47. Good, R. J., Kuikstad, J. A. and Bailey, W. O., *J. Coll. Interface Sci.*, **35** (1971), 314.

48. Schrader, M. E. in *Surface contamination–genesis, detection and control*, Vol. 2, K. L. Mittal (Ed.), Plenum, New York 1979, p. 541.
49. Parsegian, V. A., Weiss, G. H. and Schrader, M. E., *J. Coll. Interface Sci.*, **61** (1977), 356.
50. Neumann, A. W. and Tanner, W. J., *J. Coll. Interface Sci.*, **34** (1970), 1.
51. Neumann, A. W., Harnoy, Y., Stanga, D. and Rapacchietta, A. V. in *Colloid and interface science*, Vol. 3, M. Kerker (Ed.), Academic Press, London, 1976, p. 301.
52. Baszkin, A. and Ter-Minassian-Saraga, L., *Polymer*, **19** (1978), 1083.
53. Briggs, D., Rance, D. G., Kendall, C. R. and Blythe, A. R., *Polymer*, **21** (1980), 895.
54. Fox, H. W. and Zisman, W. A., *J. Coll. Sci.*, **5** (1950), 514.
55. Ellison, A. H. and Zisman, W. A., *J. Phys. Chem.*, **58** (1954), 503.
56. Kitazaki, Y. and Hata, T., *J. Adhes.*, **4** (1972), 123.
57. Schonhorn, H. and Hansen, R. H., *J. Appl. Poly. Sci.*, **11** (1967), 1461.
58. Schonhorn, H and Ryan F. W., *J. Poly. Sci.*, part A2, **7** (1969), 105.
59. Levine, M., Ilkka, G. and Weiss, P., *Polym. Lett.*, **2** (1964), 915.
60. Bernett, M. K. and Zisman, W. A., *J. Phys. Chem.*, **63** (1959), 1911.
61. Sharpe, L. H. and Schonhorn, H. in *Contact angle wettability and adhesion*, ACS Adv. Chem. Ser., **43** (1964), 189.
62. Zisman, W. A., in *Polymer science and technology*, Vol. 9A, *Adhesion science and technology*, L.-H. Lee (Ed.), Plenum, New York–London, 1975, p. 55.
63. Bewig, K. W. and Zisman, W. A., *J. Phys. Chem.*, **69** (1965), 4238.
64. Schrader, M. E., *J. Phys. Chem.*, **74** (1970), 2313.
65. Schrader, M. E., *J. Phys. Chem.*, **78** (1974), 87.
66. Huntsberger, J. R., *J. Paint Technol.*, **39** (1967), 199.
67. Huntsberger, J. R., *Adhesives Age* (December 1978), 23.
68. Fowkes, F. M., *Ind. Eng. Chem.*, **56** (12) (1964), 40.
69. Owens, D. K. and Wendt, R. C., *J. Appl. Polym. Sci.*, **13** (1969), 1741.
70. Kaelble, D. H. and Uy, K. C., *J. Adhes.*, **2** (1970), 50.
71. Owens, D. K., *J. Appl. Polym. Sci.*, **14** (1970), 1725.
72. Fowkes, F. M. and Maruchi, S., *Coatings and Plastics Preprints*, **37** (1977), 605.
73. Dann, J. R., *J. Coll. Interface Sci.*, **32** (1970), 302 and 321.
74. Krüger, R. and Potente, H., *J. Adhes.*, **11** (1980), 113.
75. Wu, S., *J. Adhes.*, **5** (1973), 39.
76. Potente, H. and Krüger, R., *Farbe u. Lack*, **84** (1978), 72.
77. Hata, T. and Kasemura, T. in *Polymer science and technology*, Vol. 12A, *Adhesion and adsorption of polymers*, L.-H. Lee (Ed.), Plenum, New York–London, 1980, p. 15.
78. Mittal, K. L., *Polym. Eng. Sci.*, **17** (1977), 467.
79. Mittal, K. L. in *Polymer science and technology*, Vol. 9A, *Adhesion science and technology*, L.-H. Lee (Ed.), Plenum, New York–London, 1975, p. 129.
80. Fowkes, F. M., *J. Adhes.*, **4** (1972), 155.
81. Fowkes, F. M. in *Polymer science and technology*, Vol. 12A, *Adhesion and adsorption of polymers*, L.-H. Lee (Ed.), Plenum, New York–London, 1980, p. 43.

82. Drago, R. S., Vogel, G. C. and Needham, T. E., *J. Amer. Chem. Soc.*, **93** (1971), 6014.
83. Drago, R. S., Parr. L. B. and Chamberlain, C. S., *J. Amer. Chem. Soc.*, **99** (1977), 3203.
84. Fowkes, F. M. and Mostafa, M. A., *Ind. Eng. Chem. Prod. Res. Dev.*, **17** (1978), 3.

Chapter 7

SURFACE TREATMENTS FOR STEEL

J. M. SYKES

University of Oxford, Oxford UK

1. INTRODUCTION

The term 'steel' covers a wide variety of alloys. For the purposes of this book it will be helpful to divide the many types into two broad classes—'steels' and 'stainless steels'. The first class includes plain carbon steels, low alloy steels and any steel which is liable to rusting when exposed without protection. The term 'stainless steel' will be used not only for iron-nickel-chromium alloys of particular compositions, but for any high alloy steels which are able to resist rusting.

The distinction between these two classes is based on the types of oxide film which form on their surfaces. The addition to certain alloying elements, notably chromium, to steel encourages the formation of thin tenacious oxide films, which render the metal 'passive' and resistant to corrosive attack. In the context of surface preparation this leads to four important differences. When hot working or heat treatment is involved in the manufacture of components, carbon steels will form a thick layer of oxide known as 'mill scale', whereas on alloy steels relatively thin protective oxide layers will be produced. If bare steel surfaces are left unprotected during manufacture or storage, stainless steel will remain in good condition, but other steels will become rusty, needing derusting before bonding. There will also be a significant difference in behaviour during chemical pretreatment for bonding; with stainless steel more aggressive acids will be needed to break down the oxide film and expose clean metal. Finally, because of their poor resistance to the environment, carbon steels are likely to suffer attack beneath adhesive or paint films, thus lowering bond durability.

Clearly caution must be exercised in using such a simple classification; for best bond strength and durability it will be necessary to optimise pretreatment to suit the particular steel.

2. MILL SCALE

When plain carbon steels are formed by hot working, reaction with the air will form a thick oxide scale which is brittle and easily damaged. In a corrosive environment any steel exposed at cracks in the oxide will suffer severe corrosion and extensive pitting will result.[1,2] This arises because the large area of oxide can act as a cathode for the corrosion reaction and form a galvanic couple with the steel. It is therefore considered good practice to remove this scale completely before application of coatings.[1-6] The formation of good adhesive bonds will also require its removal. Scales formed above 510°C have a three-layer structure. The thickest part (next to the steel) is wüstite, FeO; next comes a layer of magnetite, Fe_3O_4, then finally a thin film of Fe_2O_3.

On cooling to room temperature the inner layer disproportionates into a finely divided mixture of iron and magnetite which can be readily attacked by mineral acids. Such oxide layers are readily removed in hot sulphuric or cold hydrochloric acid,[4,6] by undermining rather than by complete dissolution. Inhibitors ('pickling restrainers') should be added to reduce attack on the metal substrate.

Mill scale can also be removed mechanically by wire brushing or chipping, but the most efficient method is grit blasting.[4,5] Chilled iron or abrasive grits are found to have a clean cutting action which not only dislodges oxide but can cut into the metal, exposing a clean surface. This is particularly important for rust-pitted surfaces, as discussed below. If shot is used, although brittle oxide can be removed, fragments may become embedded in the steel.

3. RUST

Where steel has suffered corrosion it will be covered with rust deposits composed of particles of loosely adherent hydrated iron oxides. These do not present a sound surface and must be removed for maximum adhesion.[4,7] Pickling in mineral acid will usually remove this quickly and efficiently, but care must be taken to ensure that all traces of acid are removed by thorough rinsing. Drying too must be carried out quickly and thoroughly to prevent renewed rusting. Phosphoric acid is generally to be preferred in that any residue will be less harmful, and a thin film of iron phosphate is formed which confers some measure of protection. Mechanical methods too deal speedily with rust deposits; grit blasting is preferred. Techniques such as wire brushing or abrasion which in the absence of severe corrosion will quickly give a bright finish, nevertheless do not clean the surface properly. For surfaces rusted in clean air this is not important, but in industrial or marine atmospheres the corroded surface will be contaminated with either ferrous sulphate or chloride. Crystals of these salts are to be found in pits on the metal surface and are difficult to remove. If they are left on the steel, exposure to water or a moist atmosphere will soon re-establish the corrosion process. Where paint is applied to such surfaces blistering and breakdown of the paint film are greatly accelerated. For best performance these 'salt nests' must be meticulously removed.[3,8] It is widely recognised that, despite the trouble involved in producing it, a clean 'white metal' surface (British Standard first quality,[4] Swedish Standard SA3) is essential for high performance coatings, and will repay the extra cost in terms of extended coating life. It can readily be shown that even apparently clean surfaces may still retain traces of iron salts; a 1% solution of potassium ferrocyanide applied to the steel will produce spots of blue pigment.[8] Filter papers, soaked in ferrocyanide solution and dried, can conveniently be used to check the quality of the surface; when applied to the surface and dampened these papers show a blue spot wherever there are salt deposits.

Although ferrous sulphate and ferrous chloride are both highly soluble in water, removal by water washing is not particularly effective, even after grit blasting, and can take up to eight hours.[3] Nevertheless, some experts advocate wet blasting (which may also permit the use of sand as abrasive) to increase removal of salts. It must be recalled however, particularly with paints, that the use of dry grit blasting is very advantageous in producing a dry surface. If this is coated at once the problem of displacing moisture from the steel can be avoided.

Dasgupta and Ross[9] have shown that treatment with solutions containing barium can help to render rust residues inert, converting sulphates to insoluble barium sulphate. This procedure was more successful than incorporating barium salts in the primer, though additions of barium phosphates to a zinc-rich primer have been found to improve performance on pre-rusted surfaces.[10] Presumably similarly effective procedures could be devised for chloride contamination. An ideal primer ought to perform well over rusty steel; it is interesting to note that red lead–linseed oil primer, invented over 2000 years ago, is still one of the best in this respect.[11]

In some situations, particularly in maintenance painting, it is difficult or impossible to prepare surfaces properly. Much effort has been devoted to inventing methods for converting rust to protective layers of magnetite or haematite, or of impregnating the rust with protective chemicals[9] (e.g. tannins). Though these sometimes[12] hinder the rapid coating breakdown which ensues where paint is applied to rust, their effect on adhesion is not known, and for adhesive bonding proper preparation is clearly worthwhile. Experience with coatings might, however, suggest that the use of suitable inhibitive pigments in adhesives should be helpful in preventing bond deterioration through attack on the substrates.

These difficulties can be avoided if fresh clean steel surfaces are protected against corrosion by a primer or a temporary protective such as oil or grease. Wherever possible such precautions should be taken.

4. DEGREASING

Because carbon steels corrode readily when exposed to moist air it is normal practice to coat them with oil or grease during manufacture, and cold-rolled steel strip will normally be supplied in this condition. Although some adhesives can show a certain tolerance to oils (indeed vinyl plastisols have been specially formulated for the automotive industry to be applied directly to oily steel)[13] in most cases the low energy oily surface will prevent proper wetting by the adhesive, or subsequently act as a weak boundary layer between metal and adhesive. Only if the adhesive can dissolve the oil layer without itself suffering deterioration, and bring about desorption of surface active additives from the metal surface, can successful bonding be achieved.

If organic solvents are used for degreasing, the effectiveness of the process will depend on the method employed.[7] Vapour degreasing is most efficient because each article is cleaned in fresh redistilled solvent. However, the volume of solvent which condenses on cold articles placed in the vapour depends upon their heat capacity, so that for very thin sections this method may not be particularly effective. Dipping in a tank of solvent soon leads to a build-up of grease, which will contaminate further work. Wiping too is relatively ineffective, but for large articles it may offer the most convenient method. Solvents cannot be chosen simply on the basis of degreasing power; in vapour degreasing for example, chlorinated solvents are generally used because they have low flammability, but their toxicity does present a problem and choice has to be largely a matter of compromise. Trichloroethylene is most commonly employed for vapour degreasing, but care must be exercised in its use. For dipping and wiping 1,1,1-trichloroethane or trichlorotrifluoroethane are popular.

Oil and grease may also be removed with alkaline cleaners[7] containing silicates, phosphates, carbonates or hydroxides. These are used hot by either dipping or spraying. In metal finishing the speed of these cleaners is often improved by applying a negative voltage to the work. This is termed 'cathodic degreasing', and the brisk evolution of hydrogen which takes place helps to loosen grease and soil deposits.* Ultrasonic agitation of the solution will have a similar effect.

All traces of the cleaning solution must be removed by thorough rinsing with clean water, preferably demineralised or distilled.

'Emulsion cleaners',[7] so called because they are rinsed off with water as an emulsion of grease, solvent and water, are extensively used to prepare surfaces for coatings.

5. PREPARATION OF STEEL FOR BONDING

There is much advice available on the preparation of metals for adhesive bonding.[7,14-27] Generally, provided steel is free from rust and mill scale, satisfactory joints may be produced with simple pretreatment.

The first requirement is that the liquid adhesive be able to wet the metal surface and produce good interfacial contact, with the adhesive penetrating into crevices in the surface, and all voids being eliminated from the interface. The metal will actually be covered with a thin film of oxide, which in turn is covered with a few monolayers of adsorbed water.[23] Nevertheless, this surface exhibits a high surface energy and should be readily wetted by organic liquids. If, however, the surface becomes contaminated with organic material this will present a low surface energy surface to the adhesive, which will not be properly wetted and a poor joint will be formed. Unless the contaminant layer has good mechanical properties and is able to interact strongly with both substrate

---

*Cathodic treatment should not be used with high-strength steels, where hydrogen embrittlement may ensure.[7]

and adhesive it will, in any case, act as a 'weak boundary layer', preventing good adhesion.

There are many possible sources of contamination and a wide variety of possible contaminants. Oil and grease are used as protectives and lubricants; those used in metal forming operations need to have powerful surface activity and will be difficult to remove.[24] Fingermarks will contain both grease and corrosive salts and normal laboratory air can cause rapid contamination of surfaces.[23] Even freshly prepared steel substrates, when examined by AES and XPS, were found to be contaminated with carbon-containing compounds[28] and studies have shown that 'wettability' deteriorates rapidly, especially in a humid atmosphere.[29]

If metals bearing greasy films are heat treated the organic material will become carbonised and even more difficult to remove.

Modern surface-analytical techniques can assist in recognising the nature of surface films, so that specific remedies may be applied to properly defined problems. At present their removal depends on *ad hoc* methods selected largely by trial and error. It is of interest to note, however, that when 'ultra-clean' stainless steel surfaces were prepared by argon ion etching in high vacuum, poor joints were obtained, and grit blasting and degreasing gave better performance.[30]

It is well known that paints can remove oil and grease from the substrate, especially if applied by brushing.[31] This will be most effective if there is good compatibility between the paint and the contaminant and if the paint has low viscosity. Acid groups which can react with the metal can also be helpful in displacing adsorbed molecules. Adhesives will usually have higher viscosities, but can also remove oily residues from steel; indeed, plastisol adhesives are now applied direct to oiled steel in car manufacture.[13] In general, adsorbed molecules will be removed most readily if the adhesive is cured at elevated temperature—Brown and Garnish found that heating joints briefly at the start of the cure was particulary helpful.[32]

Despite this, degreasing treatments still greatly improve the strength of joints: the more rigorous the degreasing procedures the stronger the joints. For instance, on untreated surfaces the adhesion of paint was increased by over 50% when vapour-degreased with trichloroethylene, as compared with swabbing with xylene.[31] Likewise on abraded surfaces, a three-hour extraction with methyl ethyl ketone more than doubled the adhesion obtained after simply swabbing the surface with solvent. Tests with an epoxy resin also showed the effectiveness of extraction with

methyl ethyl ketone,[32] but results using trichloroethylene were variable.
One very useful study examined the effect of degreasing with different solvents before applying an organosol to steel.[33,34] This showed that the peel adhesion of the coating varied systematically with the solubility parameter of the solvent. It increased substantially through the series cyclohexane, carbon tetrachloride, benzene, chloroform, acetone and n-pentanol, reaching a maximum at n-butanol. Further increase in solubility parameter through n-propanol, ethanol and formic acid caused a rapid decrease. The solubility parameter of liquids provides a guideline to their mutual solubilities; best solubility is found where the liquids have closely similar solubility parameters. It is conceivable then that some specific contaminant was most successfully removed by n-butanol because of their mutual compatibility[34]. An alternative explanation[33] is that traces of solvent are left adsorbed on the metal and that this produces good adhesion for coatings which have good compatibility with the residual solvent. Little attention seems to have been paid to this effect by other workers although Engel and Fitzwater[35] found that if lacquer was prepared using different solvents a similar trend of adhesion with solubility parameter was observed. This they attributed to a change in the mechanical properties of the cured lacquer coating.

Although the majority of procedures used in preparation for adhesive bonding recommend degreasing with a solvent, aqueous solutions of alkalis or detergents are also effective, and will remove both organic material and 'soil' (particulate contaminants). The use of aqueous solutions will of course necessitate a thorough drying operation before bonding.

Useful guidelines for the cleaning of metal surfaces may be found in the British Standards Institution Code of Practice CP3012[7] which specifies general cleaning techniques which are recommended as suitable preparation for adhesive bonding.

In practice the tendency of a water drop to spread on a metal surface is a good indication of surface cleanliness; if it wets the surface evenly without the tendency to form droplets, then the adhesive is also likely to wet the metal and produce a good strong bond. It should be mentioned that even where treatment produces good wetting by water, exposure of the surface to the atmosphere for a few hours produces a gradual increase in the water contact angle,[23] as contamination builds up from the atmosphere.[29] At first this does not prevent good adhesion. A small coverage with low surface energy material can have a marked effect on contact angle, even though most of the surface is still clean. As coverage increases however, adhesion will suffer and bonding should be carried out as soon after cleaning as is convenient; if long delays cannot be avoided cleaning must be repeated.

It is found that further treatments of steel are often able to produce improvements in bond strength; this may be because degreasing treatments are not completely effective in removing contamination, or because of other changes which may occur, such as an increase in roughness.

We can divide these treatments into two broad categories: chemical treatments, for instance etching, and mechanical treatments, notably abrasion and blast cleaning.

These treatments usually remove the surface oxide and often attack the metal beneath. In this way persistent adsorbants, firmly bonded to the oxide layer, are eliminated.

5.1. Surface films on steel

The normal air formed film on plain-carbon steel is composed of γ-Fe_2O_3 and is about 3 nm thick.[36] On stainless steels the oxide contains chromium, in about the same ratio as in the steel, and is much more protective.[37] Inhibitors such as chromate will encourage formation of protective passive films on steel; with chromate some chromium becomes incorporated in the film.[38] Passive films on iron consist of both Fe_3O_4 and partially hydrated Fe_2O_3 and are a few nanometres thick;[39] on stainless steels the films tend to become enriched in chromium.[40] It is doubtful whether these differences between oxides have any significant influence on initial bond strength, but they will influence joint durability.

Even a small degree of thermal oxidation at 225°C was found to cause a drastic decrease in the adhesion of PVC organosol to mild steel,[33] but an oxidation treatment which produced blade-like growths promoted adhesion of polyethylene.[41]

Chromate treatments[42] can form protective films on various metals, but are not widely used for steel. However, acid dichromate etches may produce thin chromate films and will encourage passivity in stainless steels. An inorganic primer containing chromate, which was developed for stoving paints, ('Accomet C', Albright and Wilson Ltd) has been found helpful in preparing metal surfaces for bonding.[19]

Phosphate treatments find wide use in preparing steel for painting,[43,44] not so much to increase adhesion, but to improve resistance to disbonding and corrosion at breaks in the coating. They are not usually regarded as a pretreatment for adhesives, though thin phosphate coatings can enhance bond strength appreciably[33,45] and may lead to better durability. Thick coarsely crystalline phosphate layers produce poor adhesion.[2,33,45]

5.2. Surface roughness

Both mechanical and chemical pretreatments will tend to cause some roughening of a metal surface, and indeed offer potential for producing a controlled degree of roughness. It is therefore important to identify to what degree the roughening produced by these treatments contributes to improved bond strength, to determine what constitutes the optimum surface profile (bearing in mind that this may vary from one adhesive to another, and upon the type of stresses applied) and to understand the different ways in which surface roughness influences adhesion.

Treatments such as grit blasting, which produce relatively rough surfaces (up to 0·1 mm from peak to adjacent trough)[7] are widely used for surface preparation, and are found to be effective.[14–19] Indeed, they often show significant improvement over other methods. However, when mild steel samples were ground first with coarse emery paper, then with successively finer papers, and finally polished with diamond paste, it was found that finer abrasive gave better adhesion, with the polished surface best of all.[46] It is surprising that although the rough surfaces gave weaker joints, they showed a smaller precentage of apparently interfacial failure.

On the other hand, tests with stainless steel showed that joints tested at low temperatures (where the epoxy adhesive was brittle) were stronger with sand blasted substrates than with polished, but that at higher temperatures the two treatments performed similarly.[26]

Several different effects of surface roughness can be identified—some favourable, some unfavourable. In the first place the actual area of interfacial contact will be increased (increasing the number of intermolecular bonds), but failure at the interface is rare in practical adhesive joints so that this may not, in fact, be important. A rough irregular profile can also provide 'keying' between adhesive and substrate, producing a purely mechanical bond which may be helpful under conditions, such as exposure to water, where disbonding would otherwise occur. An irregular profile will also tend to divert the failure path away from the interface into the bulk of the adhesive. The microscopic distribution of stress at a rough interface will be complex, and the bond will be subjected to both shear and tensile forces, the stress field depending upon the form of the surface profile. Under a tensile force a rough surface will act as a multiple scarf joint rather than a simple butt joint, and will in consequence be much stronger than an ideally smooth flat surface. This effect would be expected to depend upon the nature of the surface profile rather than its amplitude, so that fine-scale roughness could be just as effective as coarse.

The local stress concentrations produced within the adhesive may influence the way in which cracks initiate and propagate within the polymer; Packham[41,47] has suggested that fibres and bubbles on the substrate surface can improve peel strength because local stress concentrations increase the amount of viscoelastic deformation which occurs in the polymer during the failure process.

One problem associated with bonding to rough metal is that of obtaining proper wetting of the surface. De Bruyne[48] has considered this in detail, showing that certain types of surface profile will lead to trapping of air beneath the adhesive and poor filling of crevices. Thus too rough a surface may be undesirable. From the point of view of obtaining good adhesion for paint, if too rough a finish is used then 'rogue peaks' will be poorly covered and will soon start to rust.

6. ABRASIVE CLEANING

6.1. Grit blasting

Martin[16] suggests that abrasive cleaning is on the whole best for preparing steel surfaces, and that grit blasting is the preferred method. For grit blasting large areas, for instance to prepare large plates for painting, centrifugal machines which use rotating impeller wheels to dash the grit against the steel are the most efficient and are used to prepare freshly rolled plate for priming at the steelworks or for removing scale and rust in shipyards. In general, dry blast cleaning, in which the abrasive is entrained in a compressed air blast, will be more convenient for smaller areas, and vacuum blasting, where spent abrasive and debris are removed by vacuum extracting through a second nozzle around that through which the grit is ejected, is especially efficient. This will minimise the amount of dust produced by the process. In some instances wet blasting (using a high pressure water jet) or 'vapour blasting' (using fine abrasive slurry propelled by air or steam) may be advantageous. Wet blasting can make feasible the use of abrasives (such as sand) which would otherwise produce too great a hazard to health; it can also help to remove water-soluble impurities, especially those associated with rusting.

Guidelines for blast cleaning are presented in the Code of Practice CP 3012;[7] for mild steel, chilled iron or steel grit are suggested, but for stainless steels alumina grit (46–120 mesh) is preferred; the use of iron as abrasive would lead to contamination of the surface necessitating further cleaning or leading to discolouration by rust. Blast cleaning procedures for preparing steel for painting are given in BS 5493 (Section 2).[4] This specification suggests that fine abrasive will be more efficient for cleaning steel than a coarse one, unless heavy scale deposits are to be removed. It stresses that it is generally advantageous to aim for a smoother profile than the maximum specified in BS 4232 (0·1 mm from peak to trough). This is controlled by grit size and angle of impingement. Several standards bodies have devised standards for blast cleaning prior to painting (BS 4232,[49] ANSI/ASTM D 2200–67,[50] Swedish Standard SIS 05 59 00: 1967[51]). The British Standard[49] specifies three grades of finish: first, second and third quality, based largely on visual assessment. First quality includes the requirement that the whole surface should be free from scale, rust and other residues; second quality must be 95% clear and third quality 80% clear, with no large patches of residue. Even with third quality, good performance will be obtained from properly formulated paints under moderate conditions of exposure, but under corrosive conditions first quality is required. For adhesive bonding too a completely clean surface is needed. It is important that the grit should be clean and free from contamination, and that if compressed air is used this is clean and dry.

Shields[14] presents a comparison of bond strengths for three different steels prepared by grit blasting with No. 40 chilled iron shot or vapour blasting with a suspension of 200 grade and 400 grade garnet grit in water (a recommended procedure for stainless steels). The vapour blast treatment proved superior not only for an austenitic and a martensitic stainless steel, but for mild steel as well.

Garnish[52] found that for grit blasted stainless steel a heavy blast treatment was best for tests at room temperature, but at elevated temperature (80°C) a light blast treatment gave better adhesion. The light blast also produced better durability when bonds were exposed to water.

6.2. Abrasion

For preparing small areas simple abrasion with emery paper will remove surface films and provide good keying. If the surface is finally given a thorough degrease, perfectly satisfactory bonds will result. Silicon carbide or alumina grit from 46–120[19] mesh are regarded as suitable. 100 mesh[14] is commonly recommended for steels, but the comments above on surface roughness should be borne in mind.

It is important that debris formed by abrasion be thoroughly removed from the metal by brushing, air blast, or degreasing with solvent.[19]

6.3. Bond strength and durability

Tables 1–3 compare the performance of mechanical cleaning methods with chemical techniques. In general simple abrasive cleaning methods perform well: in one test with austenitic steel[14] they provide the highest bond strength. Garnish and Haskins[53] found that although abrasion of mild steel only gave about 10% improvement over thorough degreasing, the durability when exposed to water was substantially improved. It is however in this area of environmental resistance that chemical pretreatments can show a significant advantage (see below) and where exposure to water or a hot humid atmosphere is anticipated these methods are to be preferred. Guttmann[18] urges caution, however, in using chemical methods with carbon steels in that, unless great care is taken to ensure rapid and efficient drying after rinsing, rusting will take place.

It is clear that much of the merit of grit-blasting treatments stems from the favourable surface profile they produce; etching grit-blasted steel can lead to even higher adhesion, greater than for etching alone.[26]

7. CHEMICAL PRETREATMENTS

Chemical methods can be of value at all stages in the preparation of metal substrates. Alkaline solutions can bring about efficient removal of grease and soil, while acid pickling baths can remove oxide scales, derust surfaces and completely remove salt contamination. Etchants can remove the pre-existing oxide film and roughen the metal surface. Addition of inhibiting or oxidising species, such as chromate or phosphate, to the etching solution, or the use of separate passivation or phosphating treatments, can form a stable film on the steel with improvement of joint durability, especially during exposure to wet conditions.

There are a wide variety of different preparation methods and their relative performance will depend upon the particular steel which is to be treated and, to some extent, the nature of the adhesive to be used. Although it is possible to offer broad guidelines for a particular application a proper programme of testing with the materials in question, using test methods and conditions relevant to the service requirements, is highly desirable. Table 1 shows a comparative study with three different steels bonded with a polyvinyl–formal–phenolic adhesive under standard conditions.[14] This shows no clear ranking for the different methods, each steel needing a different preparation for best bond strength. The results in Tables 2 and 3 show clearly that in tests after exposure to water, or at different temperatures, the merit order can be changed.

In general more aggressive treatments will be needed to attack the oxide film on stainless steels and some treatments will remove oxide without etching the underlying metal. Overetching is undesirable and if treatments are not properly controlled can actually reduce adhesion.[18]

7.1. Treatments for carbon steels

A simple pickling treatment in hot dilute sulphuric acid (5–10% v/v at 60–85°C) will remove scale or rust and etch the underlying metal, or hydrochloric acid can be used at room temperature.[7] A 50% solution has been used to prepare adhesive bonds.[18] With these acids particular care must be taken in rinsing to avoid leaving corrosive residues. Etching in sulphuric acid–dichromate mixture[14] will produce a more protective oxide film.

Several authorities[14,15,18] suggest treatment in a solution of phosphoric acid in industrial methylated spirits (ethanol) which will produce a thin phosphate layer, able to protect the surface against rusting until the joint is made and enhance the resistance to corrosive environments. A 10 min treatment at 60°C can be carried out in 33–50% phosphoric acid solution.

This treatment produces deposits of 'smut' after etching, which should be removed by thorough scrubbing with a stiff nylon brush during rinsing.

A recent comparison[22] of this method with three others showed that in fact simple wire brushing could produce good bonds, as could an alkaline cleaner. However, after exposure to hot water this treatment was far superior. Another phosphoric acid etch with a solution containing iodide ion[27] ('iodophosphate') produced results which were just as good with an acrylic adhesive, but for an epoxy adhesive the phosphoric acid/ethanol treatment performed best.

Where high strength steels are prepared for bonding, methods such as cathodic treatment, or etching in acids,[7] can lead to hydrogen embrittlement of the metal.

TABLE 1

EFFECT OF PRETREATMENT ON LAP SHEAR STRENGTH OF
STEEL/POLYVINYL–FORMAL–PHENOLIC ADHESIVE

	Substrate		
	Martensitic stainless steel	Austenitic stainless steel	Mild steel
Surface pretreatment	Mean bond strength (MPa)		
(A) Grit blast with No. 40 chilled iron shot.	35·4	28·4	30·2
(B) Vapour blast with garnet grit (200 grade and 400 grade), protect with oil, vapour degrease with trichloroethylene	42·6	34·2	33·2
(C) Vapour degrease, immerse for 15 min at 65°C in 5 pbw sodium metasilicate, 9 pbw 'Empilan NP4' detergent (Marchon Products), 236 pbw water. Rinse in hot distilled water, dry at 70°C.	30·2	24·6	31·4
(D) Clean in proprietary alkaline solution.	35·6	22·2	25·9
(E) Etch 15 min at 50°C in 0.35 pbw satd. sodium dichromate soln., 10 pbw conc. H_2SO_4. Brush off 'carbon' residue, rinse, dry at 70°C.	40·0	14·88	28·2
(F) Vapour blast and etch (B and E).	42·8		
(G) Two-stage etch: 10 min at 65°C in 100 pbw conc. HC1, 20 pbw formalin solution, 4 pbw 30% hydrogen peroxide, 90 pbw water. Rinse then etch for 10 min at 65°C in 100 pbw conc. H_2SO_4, 10 pbw sodium dichromate, 30 pbw water. Rinse and dry at 70°C.	50·5	25·6	15·98
(H) Anodic etch for 90 s at 6 V in 500 g litre^{-1} H_2SO_4, rinse, dry at 70°C.	45·4	24·8	39·8
(I) Anodic etch (H), passivate in chromic acid.	46·6	26·3	38·0
(J) Etch in 10% HCl (w/v) (5 min at 50°C), rinse in 1% H_3PO_4, dry at 70°C.	25·6	0·66	2·14
(K) Etch 10 min in 10% HNO_3, 2% HF, rinse, dry at 70°C.	45·5	22·2	28·0
(L) Etch (K), passivate in chromic acid.	46·4	23·8	31·2

By courtesy of Sira Institute

TABLE 2

EFFECT OF SURFACE TREATMENT OF STAINLESS STEEL ON BOND STRENGTH
[EN 58 B STEEL–AV 1566 GB ADHESIVE (CIBA–GEIGY LTD.)]

| | Lap shear strength (MPa) | | | |
| | Initial | | After 30 days: water immersion at 40°C | |
Surface treatment	23°C	80°C	23°C	80°C
Degrease in trichloroethylene.	20·9	20·0	14·7	17·5
Degrease, light grit blast (alumina grit), degrease.	24·8	31·4	16·0	18·3
Degrease, heavy grit blast, degrease.	26·3	28·6	13·2	16·2
Etch in 100 g litre^{-1} sulphuric acid, 100 g litre^{-1} oxalic acid (15 min, 90°C), desmut by brushing.	26·2	28·9	15·1	21·7
Etch in sulphuric acid/oxalic acid as above, desmut in sulphuric acid/chromic acid.	27·3	33·9	21·7	28·8

TABLE 3

EFFECT OF PRETREATMENTS FOR STEEL ON BOND STRENGTH AND WATER
RESISTANCE [ADHESIVE: AV1566 GB (CIBA–GEIGY LTD)]

| | | (Lap shear strength MPa) | | | |
| | | Initial | | After 30 days: water immersion at 40°C | |
Material	Treatment	23°C	80°C	23°C	80°C
EN58B stainless steel	Degrease	23·9	25·6	15·7	17·7
	Grit blast	25·6	32·0	14·1	16·7
	Etch^a	27·4	35·1	27·2	29·7
EN58J stainless steel	Degrease	27·8	31·9	16·7	17·0
	Grit blast	27·3	34·0	29·0	25·0
	Etch	27·8	39·9	30·1	33·6
EN3B mild steel	Degrease	20·4	24·3	9·8	7·3
	Grit blast	23·7	27·4	15·9	18·3

^{a} 5 min etch at 60°C in 570 g litre^{-1} sulphuric acid, 100 g litre^{-1} oxalic acid.

7.2. Treatments for stainless steels

Martin[16] reports that thorough degreasing in a metasilicate–pyrophosphate solution is the best treatment for stainless steels and that etching produces no benefit. Nevertheless the chemical treatments for stainless steel are many and various. Shields[14] presents the results of a number of these (Table 1). Etching treatments produced strong bonds with martensitic steels, but grit blasting was better for austenitic steels. The best compromise treatment for the two types was anodic etching in sulphuric acid followed by dichromate passivation (I).

Smith and Hank[21] tested 19 different procedures, including many new ones, on a high strength duplex stainless steel and found that two treatments gave good strengths and durability. These were a sulphuric acid/dichromate etch (30% w/w H_2SO_4, 40% w/w $Na_2Cr_2O_7,2H_2O$; 1h at 75–80°C) and an anodic etch at a potential of about 1V against a saturated calomel electrode in 50% v/v nitric acid. The latter process has the merit of avoiding toxicity problems associated with chromate solution. Samples of these different surfaces were examined by AES, which showed that all the treatments which worked well produced chromium enrichment in the oxide film. Those which showed little chromium enrichment gave poor adhesion.

Many etching treatments leave 'smut' on the metal surface which is usually removed by scrubbing in the rinse water. Allen and Alsalim[54] observed that this process is inefficient and that much of the deposit is left trapped in crevices in the surface. Treatments with an oxidising solution such as nitric acid or chromic acid removed the smut completely and gave stronger joints. Etching in a sulphuric acid/oxalic acid mixture followed by this chemical desmutting treatment gave strong joints with good resistance to water[52] (Tables 2 and 3). An etch containing 570 g litre^{-1} sulphuric acid and 100 g litre^{-1} oxalic acid gave rapid etching (5 min) at 60°C on both EN58B and EN58J steels.

Where stainless steel has formed a heavy scale during heat treatment, then an aggressive pickling solution, e.g. nitric acid (20% w/w)/hydrofluoric acid (5% w/w)[14] will be needed; lighter scales can be removed in a ferric sulphate/hydrofluoric acid solution.[7]

8. DURABILITY OF ADHESIVE BONDS

8.1. Disbonding by water

It has long been recognised that the adhesion of paints to metal decreases rapidly when exposed to water[55] and a similar deterioration is found to occur with adhesive joints.[56] It has been proposed that some paints form a water-soluble layer at the paint–metal interface which can be attacked by water; in some cases there is evidence of such a layer. With adhesives it is reasonable to suppose that the layer next to the metal may have different properties to the bulk adhesive, such that hydrogen bonds between chains might be more susceptible to attack by water. Orman and Kerr[57] have suggested that under load 'stress hydrolysis' of bonds might occur. On the other hand water could also attack the polymer–metal oxide interface. Water can interact strongly with the oxide layer[58] and may be able to displace polar groups in the polymer which would otherwise form strong bonds across the interface. Calculations predict that whereas in dry conditions the thermodynamic work of adhesion will be positive, with water present it becomes negative so that disbonding is now energetically favoured.[59] These calculations showed that similar bond deterioration should occur in formamide, but not in ethanol, a conclusion confirmed by experiment. Ethanol causes severe deterioration in the mechanical properties of the adhesive, but not in the strength of joints. Examination of disbonded mild steel and stainless joints using surface analysis techniques confirm that the failure is indeed at the metal oxide–polymer interface.[60]

According to this analysis of the situation, the only possible remedy would be to improve the oxide–adhesive interaction so that a positive work of adhesion might be obtained even in wet conditions. Tests with silane primers[61] showed that these do bond strongly to the oxide, that water resistance is improved and that bonds no longer fail at the interface, but in the silane primer.

With aluminium substrates bond deterioration has been attributed to changes in the oxide layer beneath the adhesive, which destroy its cohesive strength.[62] It might be reasonable to suppose that mild steel, which has poor resistance to corrosion, might suffer breakdown. In bonds exposed to water the normal Fe_2O_3 film becomes converted to Fe_3O_4, but only after disbonding.[59]

Despite the thermodynamic instability of adhesion many treatments actually produce durable bonds. By what mechanism strength is maintained is not clear, but mechanical keying may play a part.

8.2. Disbonding by alkali

In metallic corrosion the usual cathodic process, reduction of oxygen, produces alkali. This leads to local increases in pH if, as is often the case, the cathodic reaction is localised. Evans[1] found that this local alkalinity tends to remove paint films from the substrate particularly around defects due to the ability of alkali to 'creep' over the metal. He proposed that formation of a film of alkaline liquid between paint and metal could lead to a lowering of the interfacial energy, especially if the alkaline solution has high affinity for both metal and paint. In many applications, notably finishes for motor vehicles, the use of phosphate treatments is most effective in minimising disbonding at scratches.[44]

8.3. Cathodic disbonding

Where cathodic protection is used to prevent corrosion at breaks in surface coatings the cathodic reduction of oxygen at regions of bare metal will generate alkali. This alkali can attack some paints, so that alkali-resistant types must be used on ships' hulls in order to avoid damage.[1] Even with high-performance coatings on pipelines, alkali may in the long term cause disbonding of the coatings.[63,64] These coatings are normally applied to grit blasted surfaces; what other treatments could enhance performance is not known, but zinc phosphate is not effective.[64]

8.4. Blistering

The integrity of the bond between metals and protective coatings can be destroyed in aqueous environments by blistering.

Early experiments showed that this phenomenon was associated with osmotic transport of water through the paint film, a blister filled with solution growing beneath the paint.[65,66] Such failures are commonly associated with salt deposits from rust, e.g. $FeSO_4$, or atmospheric contamination, e.g. $(NH_4)_2SO_4$, which have been trapped on the metal surface. Paint may also contain water-soluble species and can generate an osmotic pressure without surface contamination. The importance of osmosis to the process is demonstrated by the observation that distilled water causes more severe damage than salt or sugar solution of higher osmotic pressure. Mayne has produced blisters on glass substrates which contained no water. Their formation was attributed to stress in the paint film generated by swelling.[67] In such an instance good adhesion to the substrate may help to prevent blistering, although the presence of water will militate against high bond strength. A pressurised blister test has been used to measure adhesion for comparison with possible pressures generated by osmosis; it was concluded that the forces produced would be irresistible.[68] An epoxy coating did however show less blistering when the steel was treated to improve adhesion.[32]

On carbon steels the onset of corrosion can also cause blisters; in this case they form on cathodic areas and are filled with alkali. Meyer and Schwenk[69] found that blisters form readily during cathodic polarisation, provided that the test solution contains alkali metal cations, but not otherwise. A positively charged paint film, impermeable to cations, did not form cathodic blisters; instead blisters full of ferrous chloride were found in anodic areas.[70] Anodic blisters may form under external polarisation[69] but only in the presence of certain anions (e.g. chloride). Under a temperature gradient water vapour diffusion can also produce blisters.[69]

8.5. Corrosion
Water does not always lead to disbonding of coatings or adhesives, but if it can bring about corrosion of the adherends, then eventual failure is inevitable. The rate of transport of water and oxygen through a thick polymer coating or into an adhesive joint may be sufficiently slow to maintain corrosion at an acceptably low level, but paint coatings are too thin to be an effective barrier. If a complete breakdown is to be avoided in a corrosive environment, however moderate, the primer coat must contain an inhibitive pigment (or zinc dust) and the steel must be properly prepared.[4] In many cases, phosphate treatment[44,45] will be worthwhile, for instance with motor vehicles. These problems will decrease with increasing content of alloying elements such as nickel, chromium and molybdenum; stainless steels suffer attack only in the most corrosive conditions provided that there is sufficient oxygen to ensure passivity.

9. CONCLUSIONS

Unless surface films and contamination are removed poor adhesion is likely to be obtained. Specific chemical treatments will be particularly useful in improving environmental resistance, but results can vary widely from one steel to another and for optimum performance a test programme should be carried out for the steel in question. With stainless steel there is evidence that chromium enrichment in the oxide film enhances adhesion. Where smut forms during etching, chemical removal of the smut may be worthwhile.

REFERENCES

1. Evans, U. R., *Corrosion and oxidation of metals*, Edward Arnold, London 1960, p. 543.
2. McKelvie, A. N. and Shaw, R. E. in *The science of surface coatings*, H. W. Chatfield (Ed.), Ernest Benn, London 1962, p. 476.
3. Dasgupta, D. and Ross, T. K., *Brit. Corros. J.*, **6** (1971), 237.
4. BS 5493, 'Protection of iron and steel structures against corrosion', B.S.I., London 1977.
5. Harvey, A. A. B. in *Corrosion*, 2nd edn., L. L. Shreir (Ed.), Newnes–Butterworths, London 1976, p. 12:3.
6. Bullough, W. in *ibid.*, p. 12:16.
7. CP 3012, 'Cleaning and preparation of metal surfaces', B.S.I., London 1972.
8. Mayne, J. E. O., *J. Appl. Chem. Lond.*, **9** (1959), 673.
9. Dasgupta, D. and Ross, T. K., *Brit. Corros. J.*, **6** (1971), 241.
10. Evans, U. R. and Taylor, C. A. J., *Trans. Inst. Met. Finish.*, **39** (1962), 188.
11. Evans, U. R., *Corrosion and oxidation of metals* (2nd Supplementary Vol.), Arnold, London 1976, p. 286.
12. Bishop, R. R. and Winnett, M. A., *J. Oil Colour Chem. Ass.*, **63** (1980), 433.
13. DeFrayne, G. O. and Twiss, S. B., *Appl. Polym. Symp.*, **19** (1972), 291.
14. Shields, J., *Adhesives handbook*, 2nd edn., Butterworths, London 1976, p. 219.
15. Snogren, R. C., *Handbook of surface preparation*, Palmerton Press, New York 1974, p. 262.
16. Martin, J. T. in *Adhesion and adhesives* Vol. 2, R. Houwink and G. Salomon (Eds.), Elsevier, Amsterdam 1967, p. 87.
17. Semerdjiev, S., *Metal to metal adhesive bonding*, Business Books, London, 1970.
18. Guttmann, W. H., *Concise guide to structural adhesives*, Reinhold, New York, 1961, p. 17.
19. Anon., *Araldite bonding*, Instruction Manual A 15 k, Ciba–Geigy Plastics and Additives Co., Duxford, Cambridge, 1980.
20. BS 5350 Part A1, 'Methods of test for adhesives:adherend preparation', B.S.I., London, 1976.
21. Smith, T. and Hank, R., *Treatment of AM 355 steel for adhesive bonding*, AD–AO 74113, 1979.
22. Devine, A. T., *Adhesive bonded steel: bond durability as related to selected surface treatments*, AD–AO 53/944, 1977.
23. Thelen, E., *J. Appl. Polym. Sci.*, **6** (1962), 150.
24. Henderson, A. W. in *Aspects of adhesion*, Vol. 1, D. J. Alner (Ed.), University of London Press, London, 1965, p. 33.
25. Rogers, N. L., *Appl. Polym. Symp.*, **3** (1966), 327.
26. Jennings, C. W., *Appl. Polym. Symp.*, **19** (1972), 49.
27. Vazirani, H. N., *J. Adhes.*, **1** (1969), 222.
28. Gettings, M. and Kinloch, A. J., *Surf. Interface Anal.*, **1** (1979), 165.
29. Gledhill, R. A., Kinloch, A. J. and Shaw, J. J., *J. Adhes.*, **9** (1979), 81.
30. Baker, F. S., *J. Adhes.*, **10** (1979), 107.
31. Bullett, T. R. and Rudram, A. T. S., *J. Oil Colour Chem. Ass.*, **44** (1961), 787.

32. Brown, P. T. and Garnish, E. W., *J. Oil Colour Chem. Ass.*, **50** (1967), 331.
33. Sherlock, J. C. in *Aspects of adhesion*, Vol. 6, D. J. Alner (ED.), University of London Press, London, 1971, p. 112.
34. Sherlock, J. C. and Shreir, L. L., *Brit. Polym. J.*, **1** (1969), 34.
35. Engel, J. H. and Fitzwater, R. N. in *Adhesion and cohesion*, P. Weiss (Ed.), Elsevier, Amsterdam, 1962, p. 89.
36. Bancroft, G. M., Mayne, J. E. O. and Ridgeway, P. *Brit. Corr. J.*, **6** (1971), 119.
37. Olejford, I., *Corros. Sci.*, **15** (1975), 687.
38. Thomas, J. G. N. in *Corrosion*, 2nd edn., L. L. Shreir (Ed.), Newnes–Butterworths, London, 1976, p. 18:44.
39. Cohen, M., in *Passivity of metals*, R. P. Frankenthal and J. Kruger (Eds.), The Electrochemical Society, Princeton, 1978, p. 521.
40. Olejford, I., *Corros. Sci.*, **15** (1975), 697.
41. Evans, J. R. G. and Packham, D. E., *J. Adhes.*, **10** (1979), 177.
42. Cole, H. G. in *Corrosion*, 2nd edn., L. L. Shreir (Ed.), Newnes–Butterworths, London, 1976, p. 16:33.
43. Shaw, R. E. in *Corrosion*, 2nd edn., L. L. Shreir (Ed.), Newnes–Butterworths, London, 1976, p. 16:19.
44. BS 3189, 'Phosphate treatments of iron and steel against corrosion', B.S.I., London, 1959.
45. Schultz, J., Sehgal K. C. and Shanahan M. E. R., *Adhesion—1*, K. W. Allen (Ed.), 1977, p. 269.
46. Bullett, T. R. and Prosser, J. L., *VIIIth Fatipec Congress* (1966), 374.
47. Adam, T., Evans, J. R. G. and Packham, D. E., *J. Adhes.*, **10** (1980), 277.
48. De Bruyne, N. A., *Aero Research Technical Notes, Bulletin No. 168*, Aero Research Ltd, Cambridge, 1958.
49. BS 4232, 'Surface finish of blast cleaned steel for painting', B.S.I., London, 1967.
50. ANSI/ASTM D2200—67, 'Pictorial surface preparation standard for painting steel surfaces, ASTM, 1967.
51. Swedish Standard SIS 05 59 00, Swedish Standards Assn., 1967.
52. Garnish, E. W. in *Adhesion—2*, K. W. Allen (Ed.), Applied Science, London, 1978, p. 35.
53. Garnish, E. W. and Haskins, C. G., in *Aspects of adhesion*, Vol. 5, D. J. Alner (Ed.), University of London Press, London, 1969, p. 259.
54. Allen, K. W. and Alsalim, H. S., *J. Adhes.*, **8** (1977), 183.
55. Walker, P., *Off. Dig.*, December 1965, p. 1561.
56. MacDonald, N. C., in *Aspects of adhesion*, Vol. 5, D. J. Alner (Ed.), University of London Press, London, 1969, p. 123.
57. Orman, S. and Kerr, C., in *Aspects of adhesion*, Vol. 6, D. J. Alner (Ed.), University of London Press, London, 1971, p. 64.
58. Bolger, J. C., *Soc. Plastics Eng., Technical Paper 18*, 1972.
59. Gledhill, R. A. and Kinloch, A. J., *J. Adhes.*, **6** (1974), 315.
60. Gettings, M. and Kinloch, A. J., *J. Mat. Sci.*, **12** (1977), 2511.
61. Gettings, M., Baker, F. S. and Kinloch, A. J., *J. Appl. Polym. Sci.*, **21** (1977), 2375.

62. Noland, J. S. in *Adhesion science and technology*, Vol. 9A, L − H. Lee (Ed.), Plenum, New York, 1975, p. 413.
63. Schwenk, W., *GWF, –gas, erdgas*, **118** (1977), 7.
64. Gray, D., 'An assessment of external coatings and coating requirements for steel pipelines', *1st International Conference on the Internal and External Protection of Pipes*, Durham, 1975.
65. Kittelberger, W. W. and Elm, A. C., *Ind. Eng. Chem.*, **38** (1946), 695.
66. Kittelberger, W. W. and Elm, A. C., *Ind. Eng. Chem.*, **39** (1947), 876.
67. Mayne, J. E. O., *J. Oil Colour Chem. Ass.*, **33** (1950), 312 and 538.
68. Van der Meer-Lerk, L. A. and Heertjes, P. M., *J. Oil Colour Chem. Ass.*, **62** (1979), 256.
69. Meyer, W. and Schwenk, W., *Farbe u. Lack*, **85** (1979), 179.
70. Mayne, J. E. O., *J. Oil Colour Chem. Ass.*, **40** (1957), 183.

Chapter 8

SURFACE TREATMENTS FOR ALUMINIUM

A. C. MOLONEY

École Polytechnique Fédérale de Lausanne, Lausanne, Switzerland

1. INTRODUCTION

The adhesive bonding of aluminium has many advantages over other methods of fastening. These include weight saving, lower fabrication costs, reduced risk of corrosion and improved structural integrity. One of the main disadvantages of adhesive bonding is that surface preparation of the adherends is essential for the production of durable bonds.

Sheet aluminium for adhesive bonding has certain contaminants on the surface, including grease, mill scale, fingerprints and painted identification marks. Thus the initial step in any pretreatment procedure is their removal by organic solvents such as trichloroethylene or methyl ethyl ketone. The existing aluminium layer is unlikely to have the stability and uniformity necessary for permanent adhesive bonds. In addition strain may have been introduced into the surface by machining methods which can lower the cohesive strength.[1] As noted by McMillan *et al.*[2] an effective surface preparation method should produce an even oxide layer with the following characteristics. It must adhere strongly to the base material, and be cohesively strong, wettable by the adhesive, receptive to the formation of adhesion bonds and environmentally stable.

The pretreatment procedures which have been used for aluminium alloys prior to adhesive bonding fall into several categories, namely mechanical treatment, alkaline cleaning, chemical etching and acid anodising.

2. METHODS OF SURFACE TREATMENT FOR ALUMINIUM

2.1. Mechanical treatment

Mechanical treatments of aluminium include grit blasting, sanding and scrubbing with metal wool or scouring powder. Practical details of these treatments have been given by Snogren.[3] The original oxide is removed during the process and a new oxide layer forms immediately on exposure to air.[4] The joint strength has been found to increase with surface roughness up to a certain point; however, beyond this, contact between the adherends occurs giving discontinuities in the glue line and hence lower strength.[5] Although mechanical treatment of aluminium has been found to give quite high initial bond strengths, various studies[6,7] have shown the environmental resistance to be low unless subsequent chemical treatments or corrosion resistant primers are used.[3]

2.2. Alkaline cleaning

Various commercial alkaline cleaners are available which are generally a mixture of salts such as sodium carbonate, sodium hydroxide, sodium metaborate, sodium metasilicate, trisodium phosphate and small amounts of wetting agents (see Appendix for a typical formulation). The solutions are usually maintained at a pH of between 9 and 11 to avoid attack on the aluminium. Alkaline cleaners can be used to etch aluminium but this is undesirable for adhesive bonding.[3] If the cleaning solution contains metasilicate it is important to rinse the aluminium immediately (preferably in less than 15 seconds) because a film is formed on drying which is extremely difficult to remove.[8] Like the mechanical treatments of aluminium, alkaline cleaning results in moderately high initial strengths but poor durability.[9]

2.3. Chemical etching

The surface treatments described above are not normally used on their own but in conjunction with chemical etching. The longest established effective surface treatment for aluminium prior to adhesive bonding is the chromic–sulphuric or dichromate–sulphuric acid etching procedure.*

---

* The solutions used in the United States and in Britain differ slightly. In the United States the process was developed by the Forest Products Laboratory and has become known as the FPL etch. The solution contains sodium dichromate and sulphuric acid, having a lower concentration of acid and being used for a shorter time at a higher temperature than the solution used in Britain. The British method uses chromium trioxide and sulphuric acid (see Appendix). The surface topography produced by the two methods is very similar.[10]

During the etching process the original oxide layer is removed and approximately 1–3 μm from the metal surface is dissolved depending on the composition of the alloy. A new uniform oxide layer is then deposited. There has been debate as to whether the oxide layer is formed in the chromic acid solution[11] or in the subsequent rinsing stage, and in the final stages of etching the metal is free of oxide.[12] Recent work,[13] using Auger spectroscopy, has indicated that both dissolution of oxide and formation of new oxide take place during etching. A steady state is reached after a few minutes of etching, but a thin oxide layer is maintained on the surface. Thickening of the oxide occurs during rinsing.

There have been conflicting views concerning which oxide is formed on freshly etched aluminium.[14–16] Recent studies by Pattnaik and Meakin[17,18] using various techniques such as RHEED (reflection high energy electron diffraction), SEM (scanning electron microscopy), and TEM (transmission electron microscopy) have shown that different oxides may be formed after etching on different alloys. In addition, the structure on the top surface of the oxide and the bulk of the oxide may differ. The very thin films involved make conclusive statements about the hydration state of the oxide very difficult.[19]

In order to give uniform etching and durable bonds the solution composition of a chromic acid bath must be carefully controlled.[20] It has been found that a fresh solution does not give rise to a good surface configuration. Thus it is recommended that a new bath be artificially aged prior to use by the addition of a minimum of $1.5\,gdm^{-3}$ aluminium powder.[21] Natural ageing of a chromic acid bath used for etching copper-containing alloys will also involve some build-up of copper and this has been found to assist uniform etching.[22] Thus the addition of small quantities of copper ions is also recommended.

Conversely, various elements in the pickle solution are detrimental to uniform etching. The presence of silicates on the surface of aluminium, through inadequate rinsing after alkaline cleaning, can result in conversion to silicic acid in the pickle bath. Silicic acid has been found to interfere with etching.[23] Chlorides have been found to cause pitting during etching.[24] The presence of excess copper can result in backplating on aluminium and an upper limit of $1\,gdm^{-3}$ is recommended.[23] Similarly excess aluminium lowers the efficiency of etching and an upper limit of $18\,gdm^{-3}$ is advised.[25] In addition the presence of more than $1\,gdm^{-3}$ of iron in the bath can cause pitting of the aluminium.[25]

Considerable work has been reported on the effect of the water rinsing stage after chromic acid etching. Rinsing in deionised water, especially hot deionised water (60°C) has been found to result in low bond strengths,[4,26] whereas rinsing in tap water either at room temperature or 60°C resulted in high strengths. These results were explained by the formation of a thick, cohesively weak, oxide layer in hot deionised water but in tap water the growth of this weak oxide was inhibited by divalent anions (such as SO_4^{2-}).[27] Other workers suggested that the radius of cations present in the water affected joint strengths.[11] Cations with a radius of below 0·08 nm induced higher bond strengths which was attributed to their ability to diffuse into the oxide film. The poor bond strengths were explained as arising from the formation of a thick layer of the oxide bayerite ($\beta Al_2O_3H_2O$).[26] It has been shown subsequently[17] that the thickness of the oxide layer after etching and washing in tap water at 65°C is 22 nm (measured by ion sputtering and Auger) whereas etching and washing in hot deionised water gave an oxide layer 176 nm thick. Thus rinsing after etching should be carried out in tap water. The washing should begin immediately on removal from the chromic acid bath to prevent the conversion of Cr^{6+} to Cr^{3+} on the aluminium surface since this is deleterious to oxide stability.

The effect of various processing variables in the FPL etch procedure has been summarised in Fig. 1, which is taken from the work of Scardino and Marceau.[20] The detrimental effect of such factors as contamination of the solution by fluoride and time delay prior to rinsing are

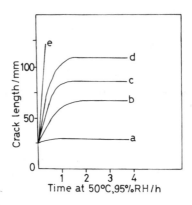

FIG. 1. The effects of FPL processing variables on the durability of wedge crack specimens bonded with a 120°C curing epoxide adhesive. (a) Correct processing; (b) new solution; (c) 5°C below normal minimum processing temperature; (d) 3 minute rinse delay; (e) contaminated solution containing 200/ppm F⁻. (After Scardino and Marceau[20].)

demonstrated by the increased crack growth in a humid environment. The oxide produced by chromic acid etching is highly reactive and a number of studies have evaluated the length of time that can be allowed between the surface treatment and priming or bonding, before degradation of the surface occurs. Wegman[28] observed a 10% drop in strength after 24 h and a further loss after 72 h exposure to a laboratory environment ($\simeq 23°C$, 50% RH). Kaelble and Dynes[29] have investigated the surface degradation after FPL etching by contact angle measurements using a range of liquids in various relative humidities. Their results indicated a regular decrease in wettability with time and they recommended a maximum of three hours' storage time in ambient conditions to minimise surface degradation. This degradation of etched aluminium has been attributed to carbon contamination. However, Smith[19] has shown, using various surface analytical tools, that degradation of the surface can occur in the absence of carbon but in the presence of water vapour. Minute concentrations of contaminants from the atmosphere can be chemisorbed on a freshly etched surface (chlorides, fluorides, carbides, etc.) giving rise to a non-wettable surface and lower strengths. Minford[30] found considerable thickening of the oxide occurred when samples were stored in a desiccator.

These investigations indicate that bonding should be carried out as soon as possible after surface treatment for maximum strength and durability to be achieved.

Danforth and Sutherland[31] have found that the surface of FPL etched aluminium can be contaminated by handling with cotton gloves or paper tissue. This was associated with carbon contamination (measured by ESCA) and was found to give rise to increased crack growth in a wedge crack test. Thus handling of the treated parts prior to bonding should be avoided.

Chromic acid etching under optimum conditions results in a surface which is covered with microscopic etch pits within larger etch pits. On the basis of evidence obtained by scanning electron microscopy, Bijlmer[32] has proposed a structure consisting of double ridged pores with anodic and cathodic areas. This was extended by Smith[19] who suggested the existence of metal ridges between the pores. In order to obtain better resolution than SEM allows, Chen et al.[33] used transmission electron microscopy to obtain stereo pairs which gave a three-dimensional view of the surface. From this they deduced a structure similar, in certain respects, to that of Bijlmer but consisting of 'whisker-like' protrusions (see Fig. 2). Chen et al. also found that incorrectly

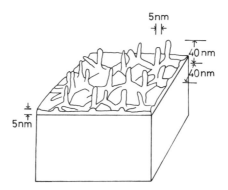

FIG. 2. Morphology of oxide produced by chromic acid etching (After Chen et al.[33]).

treated surfaces, such as those contaminated with fluoride during rinsing, did not have any protrusions on the surface.

The surface produced by chromic acid etching is highly receptive to both adhesives and contaminants. Several theories have been put forward to explain this behaviour. An early theory suggested[34] that the activity of the surface arose from the presence of unfilled d electron shells in chromium which enabled the treated surface to accept electrons from the adhesive. However, this would require the presence of larger quantities of chromium on the surface than have been observed. Several studies[35-38] have found chromium to be present at about 1% in the surface. These measurements render this theory unlikely. Smith[19] proposed that the reactivity of this surface arises from the electric field generated by the sharp metal ridges. Chen et al.[33] explain the reactivity as due to the 'whisker' morphology providing the possibility of mechanical interlocking and the formation of a large number of sites for physical forces of attraction to operate.

The oxide produced by chromic acid etching is not completely stable to moisture, which affects bond strength durability. This is clearly shown in work reported by Noland[39] using ESCA to study chromic acid etched aluminium before and after ageing for one hour at 60°C and 100% RH. The shift in the binding energy of the aluminium 2p peak position indicates a change in the oxide (see Fig. 3(i)). Noland attributed this to the conversion to a weak 'gelatinous boehmite' type oxide. More evidence is needed before definitive identification of the oxide may be made, but it is clear that some hydration of the oxide does occur.

Sun et al.[13] have provided a possible explanation for this pheno-

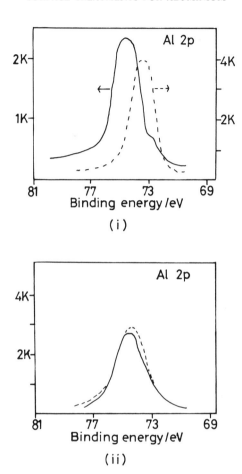

FIG. 3. (i) ESCA analysis of dichromate-sulphuric acid etched aluminium:——, unaged; --- aged for 1 h at 60°C, 100% RH. (ii) ESCA analysis of phosphoric acid anodised aluminium:——, unaged; --- aged for 1 h at 60°C, 100% RH. (After Noland,[39]).

menon. The oxide produced by chromic acid etching was studied by Auger and a copper rich layer was found to be located at the metal/oxide interface which, for certain alloys, might contain as much as 16% by weight of copper. They explained this as arising from the slow diffusion of copper, as opposed to aluminium, during the process of dissolution in etching. Baun et al.[40] have also observed a copper rich interfacial layer and have noted that changes in the acid treatment such as temperature and rinse times change the width and shape of the copper distribution as revealed by Auger spectroscopy. The presence of high levels of copper, particularly at grain boundaries, is known to assist the electrochemical dissolution of the matrix in adverse environments.[41]

Development work is under way[42] to find etching solutions without the toxicity of chromic acid. A solution of ferric sulphate in sulphuric acid[43] has been found to give strengths comparable with chromic acid etching but whether this technique can guarantee the necessary long-term durability has yet to be established. Etching in phosphoric acid has been found to give high initial strengths but poor durability.[44,45]

2.4. Acid anodising

A wide variety of electrolytes is capable of giving rise to anodic coatings on aluminium;[46] those most widely studied are sulphuric, chromic and phosphoric acids. Similar oxides are built up but their pore sizes and thicknesses, and hence their suitability to adhesive bonding, vary.

Sulphuric acid anodising of aluminium gives excellent corrosion protection for aluminium. However, problems have been encountered in its use as a surface treatment prior to bonding. It was found with peel testing at low temperatures that separation of the anodic oxide layer from the base metal occurred.[24] Failure has also been found to occur in the anodic layer rather than the adhesive.[47] A typical procedure for the sulphuric acid anodising of aluminium is given in the Appendix.

Chromic acid anodising has been used as a pretreatment process prior to adhesive bonding for many years. The processing conditions used are generally chromic acid (30–100 gdm^{-3} chromium trioxide) at 40°C for 40 min.[48] High strength bonds result with good durability.[6,14] There has been considerable debate as to whether the coating should be sealed or unsealed. Sealing has been defined as 'an aqueous treatment applied after anodising to reduce porosity and absorptivity of the coating thus enhancing the protection against corrosion'.[48] Generally, sealing has been considered to be deleterious to bond strength.[8] Bijlmer[47] found lower penetration of the adhesive into the anodic layer and subsequent fracture in this layer when the oxide had been sealed in hot water. Other workers, however, have claimed[49] that unsealed chromic acid anodic treatment gave erratic strengths and sealing in hot demineralised water containing a small quantity of chromic acid gave reliable strengths.

Chromic acid anodising produces a thick, dense oxide which consists of solid columns approximately 400 nm in diameter[50] with a smooth surface. This thick oxide improves the corrosion resistance but the lack of porosity reduces the possibility of mechanical interlocking and hence it might be expected that the initial strengths of adhesive bonds would be lower. This explanation was supported by a recent study[6] which compared chromic acid etching and anodising. The etched adherends gave higher initial bond strengths than those which had been anodised but the latter gave more durable bonds in a high humidity environment.

Although the number of variables that need to be controlled in the chromic acid anodising process are fewer than in chromic acid etching, a number of parameters still need close control which may be restrictive on a production scale.[51]

Phosphoric acid anodising as a pretreatment process prior to adhesive bonding was developed recently in the Boeing Laboratories, Seattle.[52] The process is very simple, being carried out in 10% phosphoric acid at room temperature with a potential of 10 V (see Appendix). Since the process has been found to be essentially free of processing variables, it is particularly amenable to production use. When the oxide formed by phosphoric acid anodising is viewed through a polarising filter, at low angle of incident light, complementary colours are observed at $0°$ and $90°$.[2] This test can be used in production to ensure that electrical contact was achieved during the anodising stage.

The research carried out on phosphoric acid anodising was stimulated by work published in the 1950s on the hydration of anodic coatings. Phosphoric acid anodising was found to provide[53] a more hydration resistant surface than other anodising solutions, i.e. sulphuric, chromic and oxalic acids. These acids had very similar hydration rates, but phosphoric acid was considerably slower despite the fact that the pore size was larger.[54] This work suggested that the resistance to hydration was related to the incorporation of phosphate ions into the oxide film.[53,55] During hydration of an anodic coating hydroxyl ions penetrate the porous oxide and replace the anions of the electrolyte.[56] Phosphate ions are more strongly absorbed than chromate, sulphate or oxalate and, therefore, replacement by hydroxyl ions is probably more difficult.[56,57] Thompson et al.,[58] using scanning transmission electron microscopy, energy dispersive X-ray analysis, and secondary ion mass spectrometry, have detected the presence of high levels of phosphate in the anodised structure. In addition, a gradient of phosphate concentration was observed going into the cell with almost pure alumina in the centre. Solomon et al.[59] have confirmed this observation using Auger electron spectroscopy. In addition they used Auger to study the adherend–adhesive interlayer by preparing thin film aluminium adherends and using ion sputtering to etch down to the interface. This work revealed that changes occurred during cure, the phosphate moving ahead of the adhesive into the pore vacancies.

Noland[39] has used ESCA to confirm the resistance to hydration of aluminium treated by phosphoric acid anodising; this is demonstrated in Figure 3(ii).

Venables et al.[50] have studied the surfaces of phosphoric acid anodised aluminium using transmission electron microscopy in a scanning mode. On the basis of this work they have postulated a structure for the oxide which consists of a porous columnar structure with 'whisker like' protrusions (see Fig. 4). The oxide is 200–400 nm thick on bare alloys.[21] A similar structure has been proposed by Thompson et al.[60] on the basis of evidence obtained by ultramicrotomy and transmission electron microscopy. This structure gives some explanation for the efficiency of this pretreatment procedure. If the adhesive resin penetrates the pores, an interphase is produced between the metal and the bulk of the adhesive. This provides, to a greater extent than chromic acid etching, the possibility of mechanical interlocking and the formation of a large number of sites for physical forces to operate. In addition, failure would require considerable plastic deformation of the ligaments in the pores, resulting in high fracture energies.[61]

This porous columnar structure also explains the observation that this

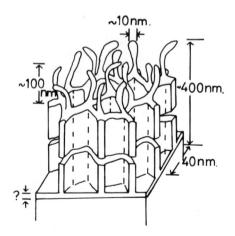

FIG. 4. Schematic representation of phosphoric acid anodised coating on aluminium (After Venables et al.[50]).

oxide can easily be mechanically damaged and thus handling should be avoided.[62]

However, for this structure to be effective in producing strong, durable bonds it is essential that the adhesive fully wets the deep pores before curing. Brewis *et al.*[63,64] have compared the durability of chromic acid etched and phosphoric acid anodised adherends bonded with two epoxide film adhesives, a 120°C curing modified epoxide adhesive BSL 312* and a 175°C curing nylon epoxide adhesive FM 1000.† The latter is well known to be moisture sensitive.[65] On exposure to a high humidity environment the anodised joints bonded with BSL 312 were more durable than the etched joints as shown in Fig. 5. In addition, for this adhesive, phosphoric acid anodising precluded the formation of patches

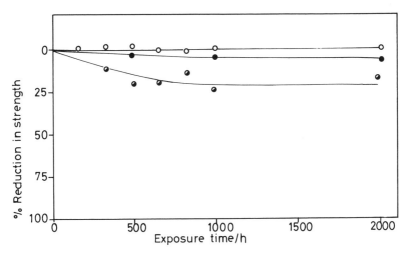

FIG. 5. The influence of surface treatment procedure on the durability of double lap joints bonded with BSL 312. ○, Chromic acid etched adherends, stored at 20°C, 45% RH; ◔, chromic acid etched adherends, stored at 50°C, 100% RH; ●, phosphoric acid anodised adherends, stored at 50°C, 100% RH.

of interfacial failure frequently observed when joints with etched adherends are subjected to moisture. However, the FM 1000 adhesive joints prepared from phosphoric acid anodised adherends showed the same strength loss as those prepared from chromic acid etched adherends, as shown in Fig. 6. When tested dry the anodised FM 1000

* Trademark of Ciba-Geigy (Bonded Structures) Ltd.
† Trademark of American Cyanamid

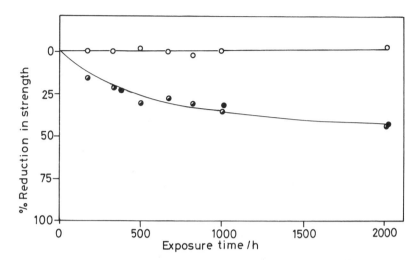

FIG. 6. The influence of surface treatment procedure on the durability of double lap joints bonded with FM 1000. ○, Chromic acid etched adherends, stored at 20°C, 45% RH; ◑, chromic acid etched adherends, stored at 50°C, 100% RH; ●, phosphoric acid anodised adherends, stored at 50°C, 100% RH.

joints showed failure in the anodic layer. The difference in the behaviour of BSL 312 and FM 1000 with respect to anodising appears to be due to the relationship between viscosity and temperature for the two adhesives. On heating BSL 312 film adhesive the viscosity of the resin decreased rapidly, the adhesive being in the molten state well below the cure temperature. On heating FM 1000, the adhesive remained in film form until 150°C when it melted and curing commenced. In fact it has been observed[66] that some nylon epoxide film adhesives will cure before the film has melted if a very slow heat-up rate is used. It is thus highly probable that curing of the FM 1000 occurred before the pores in the phosphoric acid anodised oxide had been thoroughly wetted by the adhesive. Baun[67] has provided evidence in support of this explanation, for a different adhesive system, using transmission electron microscopy.

Schwartz[68] noted that another 175°C curing epoxide film adhesive exhibited similar durability when treated by phosphoric acid anodising or chromic acid etching. This was later ascribed to the rubber based primer rather than the adhesive.[69] The porous oxide was described as acting like a molecular sieve, in that the lower molecular weight rubber molecules penetrated the pores but molecules of higher molecular weight remained at the surface of the oxide. In this manner a type of weak boundary layer was created and premature failure occurred.

The phosphoric acid anodising treatment of aluminium is a recent development. Evidence has shown that it results in a more hydration-resistant surface, which gives rise to improved durability particularly for the 120°C curing modified epoxide adhesives.[20,63] However, the morphology is such that the wetting characteristics of adhesive and primer become a factor of considerable importance.

3. OXIDE STABILITY AND JOINT STRENGTH DURABILITY

For most applications of adhesive bonding it is not the initial strength of the bonds that is of primary interest but the retention of strength in the particular service environment. The factor that has most seriously limited the use of adhesive bonding is the stability of the bonds to moisture. It has been shown that, although the mechanical properties of the cured adhesive may be restored after drying,[64] the strength of adhesive joints is only partially reversible on drying.[65,70] It appears that irreversible changes in the adherend surface are responsible. Eley and Rudham[71] have suggested that, on exposure to water, hydration of the aluminium oxide occurs, resulting in an increase in thickness due to the greater molar volume of the hydrate. On rapid drying, the hydrate decomposes, reducing the volume of the oxide film and leaving cracks. Evidence in favour of this mechanism has been found[63] in studies on adhesive joints prepared with BSL 312. Those involving chromic acid etched adherends only recovered to 83% of the original strength, whereas those treated by phosphoric acid anodising recovered completely on drying.

How does this oxide instability arise and how is it affected by the surface treatment? Some of the surface analytical tools described in Chapters 2–5 have been used to study the surface of aluminium, after certain pretreatment processes and after bonding. Some general patterns have emerged from this work which may correlate with joint strength durability.

Heat treatment of aluminium has been shown to cause segregation of magnesium to the surface of the alloy.[41] It has already been mentioned that copper can assist the electrochemical dissolution of aluminium and this is also the case for magnesium.[41] A simple solvent degreasing pretreatment does not remove this magnesium rich layer and this may be one reason for the poor durability of this method. Etching, grit blasting and anodising all remove this layer of magnesium but after cure the levels of magnesium detected on the metallic side of fractured joints were found to vary.[72] The amount of magnesium was found to increase in the order phosphoric acid anodising < chromic acid etching < grit blasting < degreasing. This correlates with the order of decreasing durability. In the case of phosphoric acid anodising Solomon and Hanlin[73] have noted the absence of alloying elements in the oxide which may explain the stability of the oxide produced by this treatment.

It has also been observed that heating at 175°C, for example during adhesive cure, causes a dramatic increase in magnesium concentration due to the diffusion of Mg^{2+} ions.[13] This factor is contrary to the role of magnesium in joint strength instability, since many 175°C curing adhesives form exceptionally durable bonds even on poorly prepared surfaces, for example, nitrile–phenolic adhesives.[21,74] Unfortunately this type of adhesive is unsuitable for modern aircraft construction due to the structural complexity.[75]

Further work is needed to clarify whether a relationship exists between magnesium concentration and joint strength durability.

4. INFLUENCE OF ALLOY COMPOSITION ON SURFACE TREATMENT

4.1. Alloying elements

The majority of studies on surface treatment of aluminium for adhesive bonding have been carried out using copper-containing alloys since these are used in aircraft construction. The recommended chromic acid surface treatment procedure has been optimised for these alloys and may not be suitable for other types of alloy.[76]

Peel specimens which were prepared according to DTD 5577,[77] using an aluminium–manganese alloy (BS 3 L61) for the flexible adherend and a copper-containing alloy BS 3 L73 for the rigid adherend, gave very low values of peel strength and, in general, interfacial failure.[78] This effect was even more pronounced when the two alloys were etched simultaneously in the pickle bath. When the BS 3 L73 alloy was used for both rigid and flexible members, much higher peel strengths were obtained with cohesive failure. The explanation of these results appears to lie in the electrochemistry of the L61 alloy. Bijlmer[23] investigated the potential differences that may be set up between two alloys which are dipped in the same etching bath and found these potential differences caused changes in etching rates.

Manganese is the major constituent of L61, after aluminium itself, and this forms MnAl$_6$ in the alloy.[79] This has the same potential as the matrix.[80] Copper, however, is cathodic with respect to the aluminium matrix. It has been reported that, during chromic acid etching of aluminium–copper alloys, localised cathodes can be set up which are equivalent to an applied potential difference of 3 V.[81] This is not possible with an aluminium–manganese alloy as the potentials are the same. These results suggest that the lack of localised cathode formation may inhibit good surface treatment. The conditions normally used for chromic acid etching appear unsatisfactory for treating aluminium–manganese alloys and this may also be the case for other alloys. In such instances, an anodising process should be used, where the applied voltage provides the driving force for the formation of a uniform oxide film.

4.2. Clad and bare alloys

The veneering or cladding of an aluminium alloy with a more anodic metal is a classical method of corrosion protection. Some aluminium alloys are clad with pure aluminium so that the resulting composite has the mechanical properties of the core alloy but is anodically protected by the cladding against corrosion, at the expense of slightly increased weight. Clad alloys have been used in aircraft construction for over 40 years and this has been accompanied by very few problems in European aircraft.[14] However, in the United States, it has been recommended that no clad alloys be used as bonding members in aircraft construction.[20,50,82] Riel[80] has proposed an anodic corrosion mechanism for the failure of joints bonded with aluminium adherends to explain the observed differences between bare and clad alloys. This mechanism is shown schematically in Fig. 7. In the case of a clad alloy, water entering the joint causes formation of electrolyte solution within the adhesive. The anodic cladding corrodes sacrificially, giving rise to delamination. As progressive debonding occurs, water can enter more rapidly into the gaps created, thus accelerating the process. Corrosion may also be accelerated at rivet and bolt holes, due to the potential differences set up. In the case of bare alloys, Riel stated that deep pits of corrosion may be formed, but not large scale delamination.

On the basis of this evidence, clad alloys are being used less in aircraft; however, when bare alloys are used for exterior surfaces, other methods of corrosion protection must be employed in order to preclude surface corrosion.[83]

The thicknesses of anodic coatings produced on bare and clad alloys have been found to differ. Phosphoric acid anodising results in an oxide

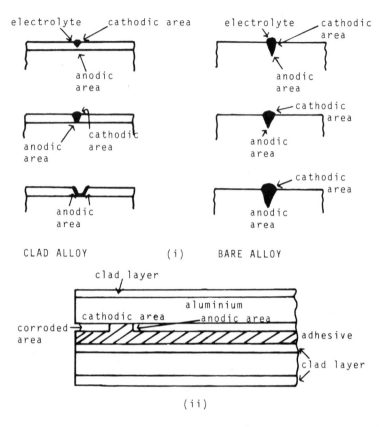

FIG. 7. (i) Progressive pitting of clad and bare aluminium alloys in a corrosive environment. (ii) Corrosive delamination of adhesive bonded clad aluminium. (After Riel[80]).

200–400 nm thick on bare alloys[21] and 500–700 nm thick on clad alloys.[84] Although chromic acid anodising produces a thicker oxide than phosphoric acid, the same pattern emerges for bare and clad alloys, as shown in Table 1.[85] SEM[85] showed that clad alloys had smoother and less porous surfaces than bare alloys.

5. PRIMERS AND ADHESION PROMOTERS

After surface treatment a primer is frequently applied to the aluminium adherend by spraying, brushing or dipping. A major function of the

TABLE 1

THICKNESSES OF OXIDES PRODUCED ON ALUMINIUM
ALLOYS ANODISED IN CHROMIC ACID (AFTER COTTER
AND KOHLER[85])

Alloy	Oxide thickness (μm)	Oxide density $(g cm^{-3})$
L 70	1·0	1·9
L 72	2·4	2·5
2024 bare	1·8	2·0
2024 clad	2·7	2·7
7075 bare	1·8	2·5
7075 clad	3·0	2·3

primer is to permit storage of the treated part since, as described earlier, the activity of the surface degrades rapidly after treatment. The primer may also assist wetting of the adherend and may serve to reduce corrosion.

Some types of primer are formulated from solvent solutions of resins which are compatible with the adhesive (for example epoxide based primers for epoxide adhesives). Corrosion inhibiting agents (such as zinc or strontium chromate) may also be added.[86] These are thought to act as oxidising agents which help to reform the protective oxide layer on aluminium when it is damaged by the ingress of water. These materials are consumed in the process and therefore have a limited use. It has been suggested[16] that these agents form $\alpha Al_2O_3H_2O$ on the interface, limiting the growth of a thick hydrated oxide.

Another type of primer or adhesion promoter are the silanes. These have the general structural formula $X_3Si(CH_2)_nY$ where $n = 0$–3, X is a hydrolysable group on silicon and Y is an organofunctional group selected to be compatible with the particular adhesive. Silanes are normally applied to the adherend as a solution to form a thin (< 100 nm) polymeric film into which the resin can diffuse.[87] Some silanes have been shown to promote increased water resistance in metal–epoxide bonds[83] and this was interpreted as being due to the formation of primary interfacial bonds between the silane and the metal. Direct evidence of this interpretation has been obtained by Gettings and Kinloch,[88,89] who used SIMS to study the interaction between γ-glycidoxypropyl trimethoxy-silane and mild and stainless steel. The existence of $FeSiO^+$ radicals for mild steel primed surface, and $FeSiO^+$ and $CrOSi^+$ radicals for the stainless steel primed surface, provided strong evidence for the existence of chemical bonding between the metal oxide and the primer. For other silane primers, which did not give rise to improved durability, the $FeSiO^+$ radicals were not detected.

Silanes form rather brittle films on metal substrates. It has been demonstrated by Auger spectroscopy and XPS[90] that, although joint durability may be improved, the silane film may become the weakest part of the joint and fracture may be initiated in this layer.

In view of the toxicity, flammability and increasing cost of the solvents used to deposit primers, work is under way to replace these organic solvents by water.[91] Water-soluble formulations based on phenol–formaldehyde, resorcinol–formaldehyde, urea–formaldehyde and special water-soluble epoxide resins have been investigated. These were applied in conventional manner and dried and cured prior to application of the adhesive. In general, the durability of these systems was not as good as conventional primers, but nevertheless further work is needed in this area. Reinhart[91] has also reported a novel process for electropriming which is of particular interest. A conducting metal tank is filled with a water suspension of the primer (approximately 10% solids) and the metal substrate made cathodic or anodic according to the charge contained by the particles of the resin under study. The possibility of forming the oxide and depositing the primer in one step was also investigated. Preliminary results obtained in this study suggest that such a methodology is feasible.

6. CONCLUSIONS

Selection of a surface treatment method necessarily involves a balance between cost and the required durability. In the aircraft industry, safety is of paramount importance, and thus phosphoric or chromic acid anodising are normally used to treat aluminium prior to adhesive bonding despite the high cost of these methods. In applications where economic factors are more important than the highest obtainable durability, mechanical treatment or alkaline cleaning may be adequate.

In the past, surface treatment procedures and adhesives have been chosen largely on an empirical basis after extensive testing. The use of the newer surface analytical tools, in particular ESCA, Auger and SIMS has provided more information on the regions a few nanometres either side of the interface. The continued application of these techniques should enable surface treatments and adhesives to be designed and chosen on a more rational basis.

ACKNOWLEDGEMENTS

The majority of this paper was prepared whilst the author was engaged in research work at Leicester Polytechnic, UK. She would like to thank her colleagues for many helpful discussions and also the Ministry of Defence, Procurement Executive, for financial support.

APPENDIX: PRETREATMENT PROCEDURES FOR ALUMINIUM FOR ADHESIVE BONDING

1. Alkaline cleaning
  (i) Vapour degrease with trichloroethylene or methyl ethyl ketone.
  (ii) Immerse in a solution containing 80 parts by weight sodium metasilicate, 15 parts by weight tetrasodium pyrophosphate and 5 parts by weight of a wetting agent such as Nacconal NR* at 70–90°C for 6–10 min. A commercial reagent such as Stripalene 532† is an alternative alkaline cleaning agent.
  (iii) Wash immediately in cold running tap water.
  (iv) Dry in warm air.

2. Chromic–sulphuric acid etch
  (i) Remove surface paint and grease with trichloroethylene or methyl ethyl ketone.
  (ii) Immerse in a mild alkaline degreasing agent, e.g. Stripalene 532 (concentration 37 g dm^{-3}) at 64°C for 5 min.
  (iii) Rinse immediately in hot tap water to remove excess Stripalene.
  (iv) Immerse in an etching solution of distilled water containing concentrated sulphuric acid (179·5 cm^3 dm^{-3}), chromium trioxide (68·5 g dm^{-3}), copper sulphate (0·39 g dm^{-3}) and stearate-free aluminium powder (5 g dm^{-3}) at 62°C for 30 min.
  (v) Wash immediately in cold running tap water for 20 min.
  (vi) Dry in warm air for 20 min.
  (vii) Bond or prime within 3 h of pretreatment.

---

* Trademark of Applied Chemical Co.
† Trademark of Sunbeam Anti Corrosives Ltd.

194   A. C. MOLONEY

3. Sulphuric acid anodising

Steps (i) to (iv) as for chromic–sulphuric acid etching.

(v) Anodise in a sulphuric acid solution containing 100–400 g dm^{-3} sulphuric acid for 20 min at 20°C with applied voltage of 14–18 V.

(vi) Wash in cold running tap water for 20 min.

(vii) Dry in warm air for 20 min.

(viii) Bond within 3 h.

4. Chromic acid anodising

Steps (i) to (iv) as for chromic–sulphuric acid etching.

(v) Immerse in a chromic acid solution containing 50 gdm^{-3} chromium trioxide at 38–40°C. Over 10 min raise the voltage to 40 V in steps of 4 V: maintain the voltage at 40 V for 40 min with a current density of 0·1 to 0·6 Adm^{-2}: increase the voltage to 50 V during the next 5 min in steps of 2 V: maintain this voltage for 5 min.

(vi) Wash in cold running tap water for 20 min.

(vii) Dry in warm air for 20 min.

(viii) Bond within 3 h.

5. Phosphoric acid anodising

Steps (i) to (v) as for chromic–sulphuric acid etching.

(vi) Anodise in a phosphoric acid solution containing 10% by weight of 85% orthophosphoric acid in distilled water. Raise voltage to 10 V in steps over 3 min and maintain full potential for 25 min. Solution temperature 20–25°C. Parts should not remain in the phosphoric acid in excess of 2 min after the power is disconnected as the oxide begins to dissolve.

(vii) Wash in cold running tap water for 20 min.

(viii) Dry in warm air for 20 min.

(ix) Verify presence of anodic coating by observation under bright light through an optical polarising filter.[2]

(x) Bond within 3 h.

REFERENCES

1. Patrick, R. L., *Treatise on adhesion and adhesives*, Vol. 4, Marcel Dekker, New York 1976.

2. McMillan, J. C., Quinlivan, J. T. and Davis, R. A., *SAMPE Quart.*, April (1976), 13.
3. Snogren, R. C. in *Handbook of surface preparation*, Palmerton Press, New York 1974, p. 118.
4. Wegman, R. I., *Adhesives Age* (January 1967), 20.
5. Semerdjiev, S., *Metal to metal adhesive bonding*, Business Books, London 1970.
6. Brewis, D. M., Comyn, J. and Tegg, J. L., *Int. J. Adhes. Adhesives* 1 (1980), 35.
7. Eickner, H. W., 'A study of methods for preparing clad 24 ST 3 aluminium alloy sheet surfaces for adhesive bonding', *Forest Products Lab. Tech. Rep. No. 1813* (1950).
8. Eickner, H. W., 'Adhesive bonding properties of various metals as affected by chemical and anodising treatments of the surface', *Forest Products Lab. Tech. Rep. No. 1842* (1954).
9. Murphy, E. B., *Nat. SAMPE Tech. Conf.*, 2 (1970), 603.
10. Bijlmer, P. F. A. and Schliekelmann, R. J., *SAMPE Quart.* (October 1973), 13.
11. McCarvill, W. T. and Bell, J. P., *J. Appl. Polym. Sci.*, 18 (1974), 343.
12. Russell, W. J. and Garnis, E. A., *SAMPE Quart.* (April 1976), 5.
13. Sun, T. S., Chen, J. M., Venables, J. D. and Hopping, R., *Applns. Surf. Sci.*, 1 (1978), 202.
14. Cotter, J. L. in *Developments in Adhesives—1*, W. C. Wake (Ed.), Applied Science, London 1977, p. 1.
15. Weber, K. E. and Johnston, G. R., *SAMPE Quart.*, 6 (1974), 16.
16. Bowen, B. B., *Nat. SAMPE Tech. Conf.*, 7 (1975), 374.
17. Pattnaik, A. and Meakin, J. D., *J. Appl. Polym. Sci., Appl. Polym. Symp.*, 32 (1977), 145.
18. Pattnaik, A. and Meakin, J. D., 'Characterisation of aluminium adherend surfaces', *Picatinny Arsenal Tech. Rep. No. 4699* (1974).
19. Smith, T., *J. Appl. Polym. Sci., Appl. Polym. Symp.*, 32 (1977), 11.
20. Scardino, W. M. and Marceau, J. A., *J. Appl. Polym. Sci., Appl. Polym. Symp.*, 32 (1977), 51.
21. Bethune, A. W., *SAMPE J.*, 11 (1975), 4.
22. Smith, A. W., *J. Electrochem. Sci.*, 120 (1973), 1551.
23. Bijlmer, P. F. A., *Metal Finishing* (December 1971), 34.
24. Reinhart, T. J., *Proc. Army Mat. Conf.*, 4 (1976), 201.
25. Bijlmer, P. F. A., *Metal Finishing*, (May 1975), 47.
26. McCarvill, W. T. and Bell, J. P., *J. Appl. Polym. Sci.*, 18 (1974), 335.
27. Wegman, R. F., Bodnar, W. M., Bodnar, M. J. and Barbarisi, M. J., *SAMPE J.* (October/November 1967), 35.
28. Wegman, R. F., *SAMPE J.* (January 1968), 19.
29. Kaelble, D. H. and Dynes, P. J., *J. Coll. Interface Sci.*, 52 (1975), 562.
30. Minford, J. D., *Adhesives Age* (March 1978), 17.
31. Danforth, M. A. and Sutherland, R. J., *J. Appl. Polym. Sci., Appl. Polym. Symp.*, 32 (1977), 201.
32. Bijlmer, P. F. A., *J. Adhes.*, 5 (1973), 319.
33. Chen, J. M., Sun, T. S. and Venables, J. D., *Nat. SAMPE Symp. Exhib.*, 22 (1978), 25.

34. Bickford, H. G., 'Treatment of metal surfaces for adhesive bonding', *W.A.D.C. Tech. Rep. No. 55–87*, Part IV (1956).
35. McDevitt, N. T., Baun, W. L. and Solomon, J. S., *J. Electrochem. Soc.*, **123** (1976), 1058.
36. Westerdahl, C. A. L. and Hall, J. R., 'Location of adhesive bond failure by using radiotracers', *Picatinny Arsenal Tech. Rep. No. 4498* (1973).
37. Packham, D. E., Bright, K. and Malpass, B. W., *J. Appl. Polym. Sci.*, **18** (1974), 3237.
38. Baun, W. L. in *Characterization of metal and polymer surfaces*, L. H. Lee (Ed.), Academic Press, New York 1977, p. 375.
39. Noland, J. S. in *Adhesion science and technology*, Vol. 9A, L. H. Lee (Ed.), Plenum Press, New York 1975, p. 413.
40. Baun, W. L., McDevitt, N. T. and Solomon, J. S., *ASTM Spec. Tech. Publ.*, **596** (1976), 86.
41. Doig, P. and Edington, J. W., *Britt. Corr. J. (Quart.)*, **9** (1974), 220.
42. Russell, W. J., *J. Appl. Polym. Sci., Appl. Polym. Symp.*, **32** (1977), 105.
43. Russell, W. J., and Garnis, E. A., US Patent Appl. 32,210, 15 February 1980, 'Low toxicity etchant for preparing aluminium alloy surfaces for adhesive bonding'.
44. Butt, R. I. and Cotter, J. L., *J. Adhes.*, **8** (1976), 11.
45. Minford, J. D., *Adhesives Age* (July 1974), 24.
46. Jenny, A. in *Anodic oxidation of aluminium and its alloys*, Griffin and Co., London 1940, p. 71.
47. Bijlmer, P. F. A., *Metal Finishing* (April 1972), 30.
48. UK Min. of Def. Std. 151, 'Anodising of aluminium and its alloys'.
49. Rogers, N. L., *J. Appl. Polym. Sci., Appl. Polym. Symp.*, **32** (1977), 37.
50. Venables, J. A., McNamara, D. K., Chen, J. M., Sun, T. S. and Hopping, R., *Nat. SAMPE Tech. Conf.*, **10** (1978) 362.
51. Shannon, R. W. and Thrall, E. W., *J. Appl. Polym. Sci., Appl. Polym. Symp.*, **32** (1977), 131.
52. Marceau, J. A., Firminhac, R. H. and Moji, Y., US Patent 4085012 (18 April 1978).
53. Hunter, M. S., Towner, P. F. and Robinson, D. L., *Proc. Amer. Electroplating Soc.*, **46** (1959). 220.
54. Keller, F., Hunter, M. S. and Robinson, D. L., *J. Electrochem. Soc.*, **100** (1953), 411.
55. Amore, C. J. and Murphy, J. F., *Metal Finishing*, **63** (1965), 50.
56. Murphy, J. F., *Plating*, **54** (1967), 1241.
57. Diggle, J. W., Downie, T. C. and Goulding, C. W., *Chem. Revs.*, **69** (1969), 365.
58 Thompson, G. E., Furneaux, R. C., Wood, G. C. and Hutchings, R., *J. Electrochem. Soc.* (September 1978), 1480.
59. Solomon, J. S. and Hanlin, D. E. in *Adhesion and adsorption of polymers*, L. H. Lee (Ed.), Plenum Press, New York 1980, p. 103.
60. Thompson, G. E., Furneaux, R. C., Wood, G. C., Richardson, J. A. and Goode, J. S., *Nature*, **272** (1978), 433.
61. Bascom, W. D., *Adhesives Age*, (April 1979), 28.

62. Thrall, E. W. and Shannon, R. W., *Adhesives Age* (July 1977), 37.
63. Brewis, D. M., Comyn, J., Cope, B. C. and Moloney, A. C., *Polym. Eng. Sci.*, **21** (1981), 797.
64. Brewis, D. M., Comyn, J., Cope, B. C. and Moloney, A. C., *Polymer*, **21** (1980), 1477.
65. Brewis, D. M., Comyn, J., Cope, B. C. and Moloney, A. C., *Polymer*, **21** (1980), 344.
66. De Lollis, N. J., *Adhesives for metals*, Industrial Press, New York 1970.
67. Baun, W. L., *Applns. Surf. Sci.*, **4** (1980), 291.
68. Schwartz, H. S., *J. Appl. Polym. Sci., Appl. Polym. Symp.*, **32** (1977), 65.
69. Marceau, J. A., *SAMPE Quart.*, (July 1978), 1
70. Kerr, C., MacDonald, N. C. and Orman, S., *Brit. Polym. J.*, **2** (1970), 67, 71.
71. Eley, D. D. and Rudham, R. in *Adhesion fundamentals and practice*, UK Min. of Tech (Ed.)., Elsevier, London 1969, p. 91.
72. Smart, N. and Kinloch, A. J., *Internat. Conf. Adhesion and Adhesives— Science, Technology and Applications*, Grey College, Durham, UK, September 1980.
73. Solomon, J. S. and Hanlin, D. E., *Applns. Surf. Sci.*, **4** (1980), 316.
74. Buck, B. I. and Hockney, M. G. D. in *Aspects of adhesion*, Vol. 7, D. J. Alner (Ed.), University of London Press, London 1973, p. 242.
75. McMillan, J. C., 'Bonded joints and preparation for bonding', *AGARD Lecture Series No. 102* (1979).
76. Thrall, E. W., *Nat. SAMPE Tech.*, **10** (1978), 73.
77. Min. of Def. Specification (1965) DTD 5577, 'Heat stable structural adhesives'.
78. Moloney, A. C., Ph.D. Thesis (CNAA), Leicester Polytechnic (1979).
79. Godard, H. P., Japson, W. B., Bothwell, M. R. and Kane, R. L., *The corrosion of light metals*, John Wiley, New York 1967.
80. Riel, F. J., *SAMPE J.* (August/September 1971) 16.
81. Smith, A. W., 'Some basic principles of surface preparation of aluminium for adhesive bonding', *Boeing Scientific Res. Labs. Doc. No. D 182–1003* (1970).
82. Reynolds, B. L., *Proc. Army Mat. Technol. Conf.* **4** (1976), 605.
83. Kinloch, A. J., *J. Adhes.*, **10** (1979), 193.
84. Remmel, T. F., 'Characterisation of surfaces prior to adhesive bonding', *Air Force Mat. Lab. Tech. Rep. AFML–TR 76–118* (1976).
85. Cotter, J. L. and Kohler, R., *Int. J. Adhes. Adhesives*, **1** (1980), 23.
86. Greer, R. H., *Nat. SAMPE Tech. Conf.*, **2** (1970), 561.
87. Bascom, W. D., *Macromolecules*, **5** (1972), 792.
88. Gettings, M. and Kinloch, A. J., *J. Mat. Sci.*, **12** (1977), 2511.
89. Gettings, M. and Kinloch, A. J., *Surf. Interface Anal.*, **1** (1979), 189.
90. Gettings, M., Baker, F. S. and Kinloch, A. J., *J. Appl. Polym. Sci.*, **21** (1977), 2375.
91. Reinhart, T. J. in *Adhesion–2* K. W. Allen (Ed.), Applied Science, London 1978, p. 87.

Chapter 9

SURFACE TREATMENTS FOR POLYOLEFINS

D. BRIGGS

ICI Petrochemicals and Plastics Division, Welwyn Garden City, UK

1. INTRODUCTION

The emergence of thermoplastics as materials with widescale potential in the packaging field, nearly 30 years ago, highlighted the distinctive nature of polyolefin surfaces. Whilst conventional ink systems could be adapted to enable plastics such as poly(vinyl chloride) and polystyrene to be printed, this was not true for polyethylene and polypropylene. At least part of the reason for this situation is to be found in Table 1 of Chapter 6. These polymers have low surface energies (and are therefore difficult to wet) and interfacial interactions occur only through weak dispersion forces. However, as will become apparent, there has been much controversy over the reason(s) for the generally poor adhesion to polyolefin surfaces.[1]

It did not take long for empirical solutions to the printing problem to be developed (a resumé of these, with the original patent numbers, is given by Gray[2]). Those most amenable to use on a large scale, especially in continuous processes, have undergone relatively little change and are the pretreatment methods used today for improving the adhesive characteristics of polyolefin surfaces. These are treatment by oxidising solutions (etchants), flame treatment and electrical ('corona') discharge treatment. Although developed originally for the pretreatment of low density polyethylene (LDPE), these methods are also used, but are not always equally as effective, for other polyolefinic materials, e.g. polypropylene (PP), propylene–ethylene copolymers, high density polyethylene (HDPE) and so-called medium density polyethylene (copolymers of ethylene and higher alkenes).

It is not the intention of this author to dwell on the practical aspects of these pretreatments. Instead, the mechanisms involved in surface modification and the reasons for subsequent adhesion enchancement will be examined in most detail.

2. ELECTRICAL DISCHARGE TREATMENT

2.1. Practical aspects

So-called 'corona' discharge treatment is the most widely used method for pretreatment of polyolefin film; it is also widely used for pretreatment of conical or cone-shaped containers (tubs). The process for treating film is shown schematically in Fig. 1. The film is passed over an earthed metal electrode which is covered with an insulating material (the 'dielectric sleeve') such as 'Hypalon', epoxide or epoxide/glass, silicone rubber or polyester. Separated from the film surface by ~ 1–2 mm is the metal electrode, usually made from aluminium. A high frequency (typically 10–20 kHz) generator and step-up transformer provides a high voltage to the electrode.

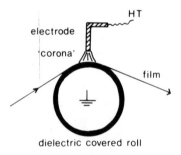

FIG. 1. Schematic diagram of electrical ('corona') discharge treatment process.

In each half-cycle the applied voltage (20 kV peak) increases until it exceeds the threshold value for electrical breakdown of the air gap, when the air is ionised and becomes plasma. The voltage eventually peaks and falls below the conducting threshold. Each cycle consists of two such events involving current flow in each direction (see Fig. 2). In continuous operation the discharge appears to be a random series of faint sparks (streamers) superimposed on a blue-purple glow (UV radiation). An atmospheric pressure plasma is called a corona discharge. Frequently, in practice, discrete intense sparks can also be seen originating from

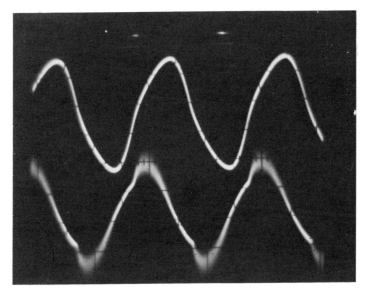

FIG. 2. Oscilloscope traces from a discharge treatment system. Upper trace: high voltage output from transformer (as applied to electrode). Lower trace: current flowing to earth, displaying periods around each half cycle maximum during which the air gap becomes conducting and discharge treatment occurs.

localised regions of the electrode, so 'corona' is not an exact description for this discharge.

The electrode–air/dielectric–metal roller combination is essentially capacitive. The oscillator/transformer circuit is inductive, so the electrical circuit as a whole can be tuned to resonance. In this condition the impedance of the generator and the treatment assembly are matched and power is coupled efficiently into the discharge. Commercial discharge treatment units attempt to achieve this condition in different ways. A great variety of electrode configurations have been proposed and used; single bar with square or rounded edge, multiple connected bars, contoured 'shoe' of larger surface area, threaded rod, etc. In general, it appears to be the case that multiple electrodes are more efficient and give more uniform treatment. Whatever system is adopted, the air gap should be as small as possible and uniform across the whole web.

The two most important parameters in discharge treatment are the power input into the discharge and the film speed. Commercial discharge treatment systems do not, in general, allow the first of these two parameters to be readily assessed. Unless impedance matching is perfect, the power input into the discharge will be an unknown fraction of the power leaving the generator and, in any case, this is not usually indicated. The method of Manley,[3] which requires the area of parallelogram generated on an oscilloscope display to be measured, is a straightforward method for obtaining the true energy dissipated in the discharge *per cycle*. The power dissipated is therefore the product of this number and the frequency.

The total discharge energy to which film is exposed is clearly inversely proportional to the linear speed. Higher line speeds can be compensated for either by increasing the energy per cycle (increasing the applied voltage) or by increasing the frequency. Before these limits are reached, however, non-uniform treatment may set in, necessitating the use of multiple electrodes or multiple treatment stations.

Containers are treated by nesting them on to a shaped earthed electrode and rotating this assembly beneath the high voltage electrode which is parallel to the treated surface. In this case the container wall is sufficiently thick to obviate the need for a dielectric sleeve.

Besides the above-mentioned primary factors which affect discharge treatment efficiency (which is used by this author to mean the energy required to achieve a given degree of surface modification, however measured), several other factors are known to be important. The constituents of the discharge gas play a role. Ozone and nitrogen oxides are formed in air discharges and their level depends, amongst other things, on the air flow rate. Introduction of nitrogen and/or carbon dioxide into an air discharge is claimed to have beneficial effects,[4] whilst high humidity lowers discharge efficiency.[5,6] Film temperature effects, both in terms of the thermal history prior to, and temperature at the point of, discharge treatment, are significant.[4,7-9] All these factors are more or less important depending on the polymer surface composition. This is a function of both the basic polymer structure and the additives contained in the material, particularly antistatic, antiblock and slip additives. The concentration of these on the polymer surface depends on temperature and time elapsed after processing. It is well known that polypropylene film is more difficult to treat than low density polyethylene; it has also been reported that PE becomes more difficult to treat as its density increases.[10] Additive bloom affects discharge treatment efficiency; when this occurs in film storage, it is found that treatment is much more difficult compared with treatment immediately after filming.[11] These factors will be discussed later.

2.2. The discharge treatment mechanism

Discussion of the mechanism by which discharge treatment modifies polymer surfaces has been intimately bound up with discussion of the reasons for improved adhesion characteristics. One particular aspect which has dominated the literature is the discharge treatment–induced enhancement of LDPE autoadhesion. This refers to the fact the LDPE surfaces can be made to autoadhere (self-adhere) after mild discharge treatment at temperatures considerably lower than normal. Until recently, two quite different theories of this effect were the cause of much controversy, which confused the basic issue of discharge treatment mechanism.

Several publications by Canadian workers[12-14] proposed that electret formation was important, the adhesion improvement being somehow related to this, although not envisaged to be simply coulombic attraction. The most compelling evidence for this theory was that the autoadhesion enhancement effect could be produced by discharge in 'inert' gases (N_2, Ar, He) as well as in air or oxygen; moreover, a given level of bond strength resulted from exposure to the same total energy of discharge, irrespective of the gas used. Also, ATR-IR spectra of treated surfaces showed evidence of chemical change only in oxidising conditions (carbonyl formation) and this change only became evident after maximum bond strength had developed.[14] The second theory, due to Owens,[15] was based on studies of film commercially treated in air. He reacted the films with various chemicals and noted the effect on autoadhesion, concluding that interfacial H-bonding between keto and enol tautomers of carbonyl groups was responsible for the enhanced adhesion.

It should perhaps be emphasised at this stage that there has been little attempt in the literature to relate commercial treatment (high frequency discharge; film speeds of the order of $1-2\,\mathrm{m\,s^{-1}}$) to laboratory treatment (variable, but often low frequency; static film). Frequently, in the latter case, only an exposure time is quoted. Let us assume an ideal case where the energy per cycle per unit area is the same for the two cases. Assume also that in the commercial process the discharge covers a 1 cm strip of film surface, that the film moves at $1\,\mathrm{m\,s^{-1}}$ and that the discharge frequency is 15 kHz. On average, every point on the film surface experiences the result of 150 cycles. For a static sample treated at mains frequency (50 Hz) the equivalent exposure time would therefore be 3 seconds. These 'ball-park' figures may help put comparative data in perspective.

Work in the author's laboratory was carried out[16] to resolve the autoadhesion controversy. Firstly, this confirms the result of Stradal and Goring,[14] that a given degree of autoadhesion enhancement results from the same total energy dissipation no matter which gas is used (in this case air, N_2 or Ar). X-ray photoelectron spectroscopy (XPS) also showed that surface oxidation resulted in all cases and that the degree, assessed by the O1s:C1s ratio, was approximately equal for similar autoadhesion enhancement. This is shown in Fig. 3. The much longer exposure times for Ar discharge result from the lower power sustainable. An improved MIR technique also showed that, in the case of air discharge, oxidation would be detected for exposure times less than required for optimum autoadhesion. Finally, this work[16] showed that discharge treatment in

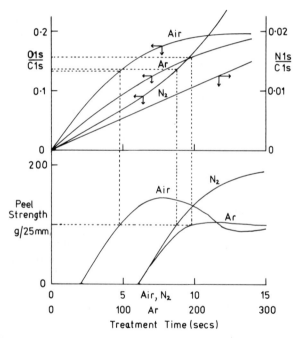

FIG. 3. Comparison of autoadhesion (peel strength) and surface composition (from XPS data) for LDPE discharge treated in air, nitrogen and argon (13·7, 13·7 and 2·2 kV respectively at 50 Hz). Heat seals were made at 75°C and 15 psi with 2 s contact time. The O1s:C1s ratio is a qualitative measure of surface oxidation level. N1s:C1s ratios refer only to surfaces treated in nitrogen. Note the similar oxidation level for samples giving peel strengths of 100g/25 mm (broken lines) (ref. 1; see also ref. 16).

H_2, with power levels in excess of those used for air discharge, did not give rise to autoadhesion enhancement or to surface oxidation (from XPS). The anomaly of the H_2 discharge had, in fact, been reported by the electret protagonists[12] without comment. Clearly these results disprove the electret theory; they also give some support to the alternative Owens theory.[15]

Typical XPS data for air discharge treated LDPE are shown in Fig. 4. As explained in Chapter 4, it is not possible to gain a detailed picture about the type and relative concentration of oxygen containing functional groups beyond the simple deconvolution shown. However, the use of derivatising agents has gone some way to overcoming this problem. Briggs and Kendall[17] subsequently positively identified carbonyl (keto-/aldehydo-), enol and carboxylic acid groups by this means. Moreover, derivatised surfaces were subjected to autoadhesion measurements. In combination these data[17,20] confirmed the Owens theory which invokes keto–enol tautomerism and hydrogen bonding between the two.

It can therefore be concluded that the main effect of discharge treatment in air is to bring about surface oxidation. It has been supposed by most workers that the most likely mechanism is free radical in nature. The corona discharge contains ions, electrons, excited neutrals (atoms and molecules) and photons. All of these can have sufficient energy to cause bond cleavage:

$$R-H \rightarrow R\cdot + H\cdot, \quad R-R' \rightarrow R\cdot + R'\cdot$$

These chain radicals react extremely rapidly with O_2:

$$R\cdot + O_2 \rightarrow RO_2\cdot \rightarrow \text{cross-links}$$
$$\Big|\ RH$$
$$\longrightarrow RO_2H \rightarrow \text{products}$$

A vast amount of work on hydroperoxide degradation reactions taking place in radiochemical, photochemical and thermochemical processes can be called upon to deduce the nature of the products. The expected functionalities include carbonyl (ketones and aldehydes), carboxylic acid and ester, alcohol and ether. Chain scission is involved in the formation of many of these groups, leading to a progressive reduction in the average molecular weight and finally to the production of CO, CO_2 and H_2O. Besides this oxidative degradation, there will be direct degradation via ion-induced sputtering. Radicals can be formed relatively deep in the polymer by UV photons. In crystalline regions these may lead to

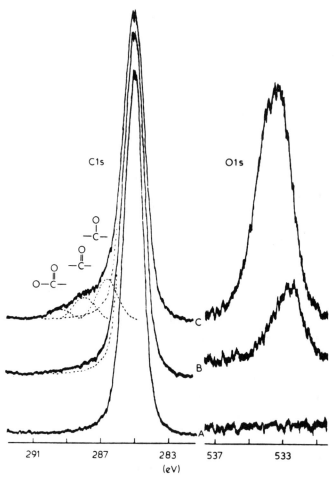

FIG. 4. High resolution C1s and O1s spectra from LDPE: (a) untreated, (b) and (c) treated in air (13·7 kV peaks voltage at 50 Hz) for 8 s and 30 s respectively. Count rates are 3×10^3 counts s^{-1} fsd (C1s) and 10^3 counts s^{-1} (O1s).[16]

unsaturation; in amorphous regions cross-linking reactions are possible. The amorphous regions are more reactive in general because of their much greater accessibility to diffusing O_2. Further reactions can also take place with ozone and oxides of nitrogen (and nitric acid), especially when low gas renewal rates are involved. Spectroscopic evidence has been provided for carbonyl (ketone or aldehyde),[16,17] carboxylic acid[17] and ester,[18] ozonide/ether,[16] nitrate ester,[16,19] nitrite[19] and -OH.[10] The key intermediate, hydroperoxide, has also recently been detected[20] (by XPS) through its reaction with SO_2 to form alkyl hydrogen sulphate (OSO_2OH).[21] There is a reasonable body of evidence to suggest, therefore, that this mechanism is involved.

In the case of 'inert' gases, it is generally believed that oxidation follows radical formation (as above) once the polymer is exposed to air, although only traces of oxygen are required during discharge treatment to effect the oxidation of the surface. In nitrogen discharge treatment, however, a high concentration of nitrogen functions is also found; these have been detected by XPS[16,22] and are thought to be mainly $-NH_2$ groups.

The free radical mechanism outlined above is consistent with most experimental observations. Firstly, and by contrast to thermal oxidation, antioxidants do not affect discharge treatment efficiency.[23] This is easily understood, since radical formation is not dependent on the chain oxidation reaction. Secondly, when extrapolating data from other free radical oxidation studies,[24] the rate of degradation should be in the order PP > propylene–ethylene copolymers > PE. It is certainly true that PP degrades faster than PE during discharge treatment.[25] The ease of treatment might reasonably be expected to follow the reverse trend. That this is so is confirmed[26] by Fig. 5. Thirdly, polymer morphology effects should be significant. This is demonstrated by two papers[10,27] showing

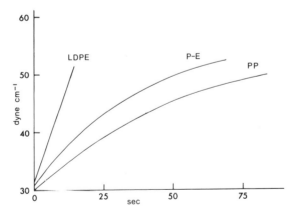

FIG. 5. Change in surface energy as a function of discharge treatment time in air for LDPE, PP and a propylene–ethylene copolymer (P–E). Surface energy was assessed using the ASTM wipe test.

that increasing the density (crystallinity) of PE makes discharge treatment more difficult, and also by a report[9] which shows that treatment of cast polypropylene copolymers becomes more efficient as the film cooling rate increases, i.e. as the crystallinity decreases.

2.3. Adhesion improvement by discharge treatment

Discharge treatment of polyolefin film surfaces is essential for the adequate adhesion of printing inks, barrier coats, vacuum deposited metals, adhesives and other polymer films used in lamination. Although there has been much speculation about the reasons for poor adhesion in the absence of discharge treatment, there has been little study of the adhesion mechanisms involved between discharge treated polyolefin surfaces and any of these adherends.

Briggs and Kendall[17] have provided some insight into the adhesive interaction between discharge treated LDPE and a commercial (nitrocellulose-based) printing ink. Using reagents which 'block' specific functional groups, they showed that if enolic —OH was eliminated from the discharge treated surface, then ink adhesion, assessed by the 'Scotch tape test',[28] dropped to zero. On the other hand, elimination of carboxylic acid functions brought about only a slight loss of ink adhesion. In none of the experiments did exposure to the relevant solvents alone affect ink adhesion. Those surfaces which suffered a loss of printability did not show a significant change in surface energy. By analogy with the autoadhesion findings (see above), it was concluded[17] that hydrogen bonding between enolic—OH groups and a component (or components) of the printing ink (probably a polyamide) was mainly responsible for the adhesion. Studies of this type have great potential in studying mechanisms of adhesion to pretreated polyolefin surfaces.

An important consideration is obviously the degree of surface oxidation, both in terms of the number of oxygen-containing functions and their depth distribution. Clark and Dilks[29] show that the O1s:C1s intensity ratio (from XPS) for discharge treated PE (of unspecified origin) is invariant to change in 'take-off angle'. This is consistent with other observations[30] on commercially treated film (having high surface energy). However, for treatment under controlled conditions, it has been found[20] that the O1s:C1s ratio is sensitive to take-off angle in the early stages of treatment. Under these conditions treatment is thus within $\sim 30\,\text{Å}$ of the surface.

From the studies of discharge treated LDPE surfaces[17] XPS showed that the attainment of excellent ink adhesion corresponded to an O:C atomic ratio of $\sim 4\%$. However, from the derivatisation results it can be estimated[20,31] that the ink adhesion is due to only $\sim 0.4\%$ of the surface carbon atoms being converted into potential enolic —OH groups. This represents a concentration of 2×10^{12} groups cm^{-2} (assuming 5×10^{14} carbon atoms cm^{-2} in LDPE film).

2.4. Other effects of discharge treatment

2.4.1. Heat sealing
Although discharge treatment can enhance the heat sealing properties of LDPE, as discussed above in the case of the autoadhesion effect, the level of treatment required to bring this about is relatively low. With the higher levels of treatment required for polyolefin film conversion it is often found that heat sealing which involves even one treated surface is impaired, i.e. at a given sealing temperature the peel strength achieved is lower or, alternatively, higher temperatures are required to achieve a given peel strength compared with untreated surfaces. Bartusch[32] has discussed the phenomenon, in the case of LDPE, in terms of changes in molecular weight distribution, crystallite size and chain mobility. It is most likely that the heat sealing is impaired by the presence of low molecular weight polar material at the interface. Support for this view comes from the fact that the removal of loosely attached material by solvent dipping restores much of the lost peel strength.[33]

2.4.2. Reverse-side treatment
This effect is caused by the creation of a region of electric discharge which impinges on the 'reverse' side of the film being treated. This can arise particularly with unslit tubular film ('lay-flat') in the edge-fold region. Other causes are poor wrap on the dielectric roll, creases in the film and wrinkles or poor jointing in the dielectric sleeve. All these allow the trapping, between the film and the dielectric roll, of air in which a discharge can be set up. Reverse-side treatment causes non-uniform treatment and can lead to ink pick-off when printed reels are unwound, and also to severe blocking (self adhesion) of reels during storage.

2.4.3. Additive effects
Any low molecular weight material which can migrate to the polymer surface before treatment (this includes the low molecular weight tail of the polymer itself) is a potential hindrance to discharge treatment. Greater energy dissipation is required to achieve the desired level of

treatment because the overlayer has to be removed (through oxidation, volatilisation, sputtering, etc). Slip and antistatic agents obviously play a strong role here. Treatment immediately after processing, before additive migration at the surface can occur, is the answer to this problem.[11]

Discharge treatment also affects slip and antistatic performance. Especially in the case of films with low levels of slip additive, the treated side has inferior slip properties (higher coefficient of friction) compared with the untreated side.[28] Willis and Zichy[19] concluded from MIR data that the slip agent oleamide exists in the crystalline form on an untreated PP surface but in an amorphous (in effect dissolved) state in the discharge treated surface layer. Antistatic properties are actually enhanced by discharge treatment.[28] Since antistatic agents perform better the higher the relative humidity, they presumably increase surface conductivity through increased moisture adsorption. The more polar treated surface will enhance this effect, but it may also lead to a higher surface concentration of the antistatic additive.

2.5. Glow discharge treatment

In this variant of electrical discharge treatment, also simply referred to as 'plasma treatment', the plasma is formed in a gas at low pressure (typically 1 Torr). Although electrode systems have been used, electrodeless systems are more convenient. These are of two types: radio-frequency discharges using either capacitive or inductive coupling and microwave discharges. A great deal of research has been carried out in this field in recent years, especially into plasma polymerisation, wherein an organic monomer is polymerised onto a substrate placed within or close to the plasma region. Since the scale-up of these techniques to a size suitable for industrial production still seems to be problematical, any further discussion is outside the scope of this chapter. A very useful source of information on this subject is the book edited by Hollahan and Bell.[34]

3. FLAME TREATMENT

3.1. Practical aspects

Although flame treatment can be, and has been, applied to polyolefin films, it is now predominantly used for pretreating articles of thicker section, particularly blow moulded bottles.

A diagrammatic representation of the set-up for flame treatment of a

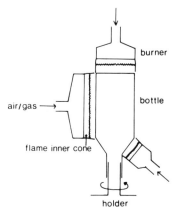

FIG. 6. Schematic diagram of set-up for flame treatment of blow moulded bottles.

standard bottle shape is given in Fig. 6. Each burner consists of a large number of closely spaced jets, usually in two staggered rows. By inclining these correctly the whole surface can be treated. The bottle passes through the fixed burner geometry and rotates at *fairly low speed* at the same time. The burners are fed with an air–gas mixture in controlled proportions. The gas can be mains (largely methane) or bottled (propane or butane). Typically the article spends about 1 s in the flame.

The most important variables in the process are the air:gas ratio; the air/gas flow rate; the nature of the gas; the burner–surface separation and the exposure time.

It is generally agreed[35–37] that for optimum treatment an oxidising flame should be used, i.e. with an excess of oxygen over that required for complete combustion. An excess of $\sim 10\%$ is recommended. Thus for methane the optimum air:gas ratio (v/v) would be approximately 11:1. Ayres and Shofner[37] found an optimum exposure time for an oxidising methane flame (unspecified polyolefin) of 0·02 s and under these conditions the optimum distance from the polyolefin surface to the tip of the blue inner cone of the flame was ~ 9.5 mm. Hurst and Schanzle,[35] however, conclude that the optimum treatment time should be ~ 0.1 s and the surface–cone tip distance should be ~ 6.4 mm. They also give a length for the inner cone of 6·3–19·0 mm. Other conditions being optimal, the nature of the gas and the flame temperature are not important.[37] For a given air:gas ratio, treatment level in a given exposure time increases as the volume of mixture burned increases.[36,38]

Most of the above discussion has concentrated on optimum conditions. These are appropriate to achieving treatment at the highest possible line speeds and, in fact, much of this work was aimed at film treatment. The above data indicate that flame treatment can easily be overdone; some of the optima in process parameters are quite sharp.[36,37]

In the treatment of containers, line speeds may not be high enough to cope with optimum flame conditions. This is especially likely when treatment time is not an independent variable (e.g. in multistage machines such as those which thermoform, treat and print in line). In these circumstances the aim is to achieve an acceptable level of treatment under variable conditions (changing line speed, polymer composition, ink used, etc). The easily adjustable parameters are air:gas ratio and flow rate, and in many cases adjustment of the gas flow rate only is sufficient. Effective treatment can be achieved, provided the exposure time is sufficiently long, at air:gas ratios much less than stoichiometric.[38]

3.2. Additive effects
Resin formulations for articles likely to be flame treated are simpler than film formulations. Apart from antioxidant, the most usual additive is an antistatic agent. Antioxidant has been shown not to affect flame treatment efficiency in the case of 0·02%di-t-butyl-p-cresol added to LDPE.[38] As with discharge treatment, flame treatment immediately after formation of the article is best. Some, but not all, antistatic agents adversely affect flame treatment efficiency when they have migrated to the polyolefin surface. Antistatic performance after flame treatment depends on the particular agent used. Performance can be better or worse than without flame treatment and this may be related to the claim that flame treatment can lead to a significant drop in permeability of polyethylene[39] as well as to the effect discussed in section 2.4.3.

3.3. The effect of flame treatment
In contrast to the large amount of work which has been carried out on discharge treatment, the flame treatment of polyolefins has received little attention.

Briggs and co-workers have recently reported[38] an XPS study of flame treated LDPE. This showed that, as is commonly assumed, the treatment brings about surface oxidation. Typical spectra are shown in Fig. 7, whilst Table 1 summarises the results of the investigation, including

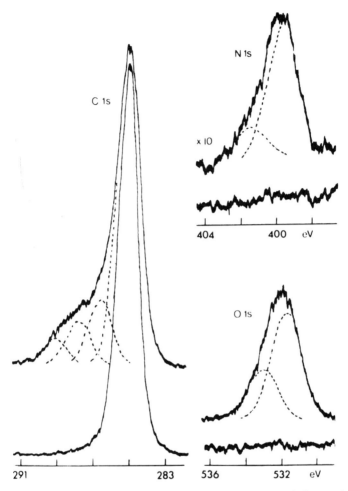

FIG. 7. High resolution C1s, O1s and N1s spectra from LDPE before and after flame treatment (4·8 s at 290 rpm with air and natural gas flow rates of 150 and 37 cm³ s⁻¹ respectively). Count rate = 3 × 10³ counts s⁻¹ fsd.[38]

measurements of adhesion to an epoxide adhesive. These data clearly show that for a fixed air:gas ratio (4·1–4·4, less than 50% stoichiometric) the surface composition depends on the volume of mixture burned, the degree of modification being lower at the lowest flow rate. Oxidation increases with time spent in the flame.

TABLE 1
XPS AND JOINT STRENGTH DATA FOR POLYETHYLENES WHICH HAVE BEEN FLAME TREATED

Polymer	Treatment time (s)	Natural gas flow (cm³ s⁻¹)	Air flow (cm³ s⁻¹)	O:C (at. %)	N:C (at. %)	Lap shear[c] strength (MN m⁻²)	Standard deviation σ sample
Alkathene 47[a]	0	0	0	0·25	0	0·55	0·07
Alkathene 47	1·2	37	150	16·9	0·94	6·6	0·6
Alkathene 47	4·8	37	150	31·0	3·2	7·2	0·7
Alkathene 47	1·2	74	317	15·3	2·2	6·8	0·7
Alkathene 47	1·2	18·5	75	6·8	0	5·1	0·6
Alkathene 11[b]	0	0	0	< 0·25	0	0·36	0·04
Alkathene 11	1·2	37	150	20·5	1·5	5·6	0·5
Alkathene 11	4·8	37	150	33·4	3·2	7·2	0·4
Alkathene 11	1·2	74	317	13·7	2·5	6·4	0·3
Alkathene 11	1·2	18·5	75	5·1	0	5·7	0·4

[a] Contains no additives.
[b] Contains 0·02% 2,6-di-t-butyl-p-cresol.
[c] With the treated polymers the failure was always a mixture of apparent interfacial and material.
Reproduced from reference 38 with permission.

It can be seen from the spectra that the overall type of surface oxidation is similar to that achieved by discharge treatment (cf. Fig. 4). The nitrogen function have BEs consistent with —NH_2 or —CN (399·7 eV) and possibly —CONH— ($\sim$401·5 eV). Angular variation experiments showed no change in the O1s:C1s ratio even for the least oxidised surface, so the modification depth is greater than 40 Å. However, measurement of the O1s:O2s ratio gave in all cases values between 16 and 20, indicating the modification depth to be less than 90 Å. The interesting result is therefore that, even for a surface with more than 30 at.% oxygen, the treatment depth is between 40 and 90 Å.

The data in Table 1 also clearly show that antioxidant, at least at the 0·02% level, does not affect flame treatment efficiency. Re-examination of surfaces after more than 12 months showed no striking changes and adhesion was not impaired. For all the samples the adhesion to an epoxide was very good, indicating that, even with the flame conditions being far from optimum, a 1·2 s treatment gave acceptable results.

3.4. The flame treatment mechanism

There are several possibilities for the mechanism of surface modification, with very few data to help choose between them. During its passage through the flame, the polyolefin surface can experience ambient temperatures of around 2000°C, so that an accelerated form of thermal oxidation could take place. This is a chain-reaction free radical process which should be prevented by antioxidant. There is some evidence,[40] however, that even at 300°C the effectiveness of antioxidant is reduced by volatilisation, so, within the thin surface layer which is oxidised, it is easy to envisage loss of antioxidant during flame treatment. Nitrogen species are also introduced into LDPE surfaces during thermal oxidation.[40]

Flames are also plasmas, characterised[34] by electron densities of $\sim 10^8$ cm^{-3} and electron energies of ~ 0.5 eV (cf. electrical discharge types of plasma with corresponding figures of $\sim 10^{11}$ cm^{-3} and ~ 5 eV respectively). Thus many excited species are present in the flame, namely free radicals, ions, excited neutrals (atoms and molecules) and electrons. Excited species which have been observed include O, OH, NH, NO and CN.[41] Incorporation of oxygen and nitrogen species into the surface, of all types observed, would arise from free radical oxidation processes of either the 'thermal' or 'plasma' types and by direct addition of radicals of the above varieties. At this stage, it is not possible to decide whether one process dominates or whether the oxidation takes place by a combination of these processes.

4. CHEMICAL TREATMENTS

4.1. Practical aspects

When the pretreatment of irregular, and particularly large, articles is required, chemical treatments involving immersion in a solution (etchant) must be used when flame treatment is not convenient. The pretreatment of polypropylene mouldings for metallisation is a familiar example.

Several formulations for etching solutions have been described. Gray[2] reports a potassium permanganate 'dip' consisting of a saturated solution of the salt in sulphuric acid ($H_2SO_4:H_2O = 1:40$). More common are chromic acid solutions, e.g.

(a) CrO_3/H_2O (900 g litre^{-1})[42]
(b) $K_2Cr_2O_7/H_2O/H_2SO_4$ (7:12:150 by weight)[43]
(c) $K_2Cr_2O_7/H_2O/H_2SO_4$ (5g/12·5 ml/87·5 ml)[2]
(d) $CrO_3/H_2O/H_2SO_4/H_3PO_4$ (2·24:15·66:42·70:39·40 wt%)[44]

Etching is usually carried out at elevated temperatures (typically 60–80°C) for several minutes. If the temperature is too high overtreatment can result; that is, a layer of highly oxidised polymer is formed, which, on subsequent adhesion testing, can fail cohesively at low applied force. This seems less likely to happen if etching is carried out at lower temperatures but for longer times.[37] Another problem with high temperature etching is distortion of the moulding.

A further pretreatment which has received some attention is oxidation by peroxidisulphate ($S_2O_8^{2-}$) solutions. Morris had described[45] the pretreatment of LDPE and HDPE by this method.

The various chromic acid pretreatment processes have received most attention in the literature and the following discussion will concentrate on them.

4.2. The effects of chromic acid

Compared with electrical discharge or flame treatments, the most obvious distinguishing feature of chromic acid treatment is the effect on surface topography. Blais et al.[46] have shown that for LDPE and HDPE film transmission electron microscopy reveals evidence of severe roughening by chromic-sulphuric acid ($K_2Cr_2O_7/H_2SO_4/H_2O$). The effect on homopolymer PP depends strongly on thermal history. Scanning electron microscopy (SEM) has shown that PP sheet with well-developed spherulitic structure is selectively etched in the regions of low or zero crystallinity. SEM pictures given by Perrins and co-workers[44,47] show the rapid roughening of propylene copolymer mouldings (by $CrO_3/H_2SO_4/H_3PO_4/H_2O$). They also present evidence for selective attack at spherulitic boundaries.

Clearly, chromic acid 'etches' polyolefin surfaces and results in material loss into solution. Blais et al.[46] found etch rates to follow the trend PP > LDPE > > HDPE. Orientation of PP film reduced the etch rate to below that of LDPE. They also found etch rates to be independent of time up to 6 h at 70°C. Similar results have been obtained by McGregor and Perrins[47] for copolymers and by Ghorashi[42] for PP homopolymer.

Water contact angles on LDPE and HDPE fall rapidly as etching proceeds, but in the case of PP a minimum is reached after a short time followed by an increase to a plateau value similar to that of the untreated surface.[43,46]

Reflection IR studies by Blais et al.[46] revealed extensive chemical changes to the surface in the case of LDPE but not HDPE or PP. They detected new bands corresponding to the introduction of $-OH, > C=O$ and possibly $-SO_3H$ groups. Later, higher quality, spectra by Willis and Zichy[19] confirmed these results for LDPE. XPS studies of similarly treated LDPE and PP by Briggs et al.[43] showed oxidation and sulphonation to have taken place in both cases with evidence for

$$C-OH, \quad \diagdown C=O, \quad -\overset{\overset{\displaystyle O}{\|}}{C}-O \quad \text{and} \quad -SO_3H \text{ groups}$$

Contact angles were also reported, and these confirmed the trends noted by Blais et al. Several types of information from XPS[43] (O1s:C1s ratios as a function of etching time and of take-off angle; O1s:O2s ratios) showed that the surface oxidation of PP rapidly (e.g. after 1 min at 20°C) reaches an equilibrium state and involves only a thin layer of the polymer (< 100 Å), whereas LDPE oxidation increases in degree and in depth as a function of time. These investigations have recently been extended[48] to etching with chromic–sulphuric acid under very mild conditions, with concentrated sulphuric acid and with CrO_3/H_2O, and include data for HDPE. These results confirm the hypothesis of Blais et al.,[46] that rapid attack at tertiary carbon atoms results in material loss of oxidised PP into solution at an early stage in the reaction compared with LDPE, the equilibrium oxidised layer thickness being insufficient for detection by reflection IR. They also show that as surface roughness increases, little can be inferred about surface polarity from contact angle measurements.

D. BRIGGS

Table 2 collects some of these XPS data which have been quantified,[43] to a limited extent, assuming that all surface sulphur is in the form of SO_3H groups. Also included are adhesion data for these surfaces with an epoxide adhesive. Comparison of XPS and adhesion results shows that there is a very good correlation between adhesion level and degree of surface oxidation (expressed in terms of the number of carbon–oxygen functions with an assumed constant average stoichiometry), but no correlation between adhesion and $—SO_3H$ group concentration.

A similar correlation was found[48] for CrO_3–H_2O etching of LDPE as shown in Fig. 8. This also shows that firmly bound Cr is present on the surface after etching but the surface concentration passes through a sharp maximum. That this residual Cr is important to achieving optimum adhesion of electroplated metal adhesion has been demonstrated by Perrins and Pettett[44] and by Ghorashi.[42] They showed that it could be removed by acid treatment and XPS confirms this. XPS spectra[48] for the O1s and C1s levels as a function of etching time with CrO_3–H_2O are shown in Fig. 9. The O1s BE maximum and the form of the C1s shoulder indicate that $>C{=}O$ groups dominate in the early stages of the oxidation process but carboxylate (acid and ester) dominate in the later stages. Reflection IR confirmed this interpretation. The loss of Cr from the surface corresponds to the transition point.

The precise chemical state of a polyolefin surface after chromic acid etching is therefore dependent on the nature of the polymer, its thermal history, the composition of the etchant solution and the time and temperature of exposure.

4.3. The mechanism of chromic acid oxidation
Little work has been reported on this subject. By analogy with the postulated reaction mechanisms for the oxidation of aliphatic hydrocarbons it can be assumed that the following sequence is likely:[49]

$$R_3C—H + H_2CrO_4 \rightarrow R_3C—OH + Cr(IV)$$
$$R_2CH_2 + H_2CrO_4 \rightarrow R_2CHOH + Cr(IV)$$
$$Cr(IV) + Cr(VI) \rightarrow 2Cr(V)$$
$$R_2CHOH + Cr(VI)/Cr(V) \rightarrow R_2C{=}O + Cr(IV)/Cr(III), \text{etc}$$

The reactive species is thought to be $HCrO_4^-$ and the reactivity of C—H bonds is in the order $R_3C—H > > R_2CH_2 > R—CH_3$. Further oxidation of $R_3C—OH$ and $R_2C{=}O$ takes place to give carboxylic acids, involving chain scission but the mechanisms are obscure. The final stable state of Cr is Cr(III). The XPS data of Briggs et al.[48] is consistent with

TABLE 2
SURFACE COMPOSITION AND ADHESION OF ETCHED POLYMERS

Polymer	Etching conditions	C:S atomic ratio	O:S atomic ratio	C atoms with SO_3H groups (%)	O (not in SO_3H groups) to total C (% O:C)	Lap shear strength ($MN\,m^{-2}$)	Failure[b] type	O1s:O2s[c]
LDPE	None	—	—	—	0·25	0·55	I	—
LDPE	Conc. H_2SO_4 1h at 70°C	36·8	3·2	2·7	0·6	3·3	I	11·6
LDPE	Acid B[a] 5 s at 20°C	198	9·1	0·5	3·1	4·8	I+M	—
LDPE	Acid A[a] 1 min at 20°C	269	12·7	0·4	3·6	7·5	M	13·0
LDPE	Acid A 30 min at 70°C	80·0	10·4	1·3	9·3	7·6	M	9·2
LDPE	Acid A 6 h at 70°C	47·1	9·5	2·1	13·9	9·5	M	9·9
HDPE	None	—	—	—	0·52	0·38	I	—
HDPE	Conc. H_2SO_4 1h at 70°C	64·3	4·2	1·6	1·8	3·5	I	18·2
HDPE	Acid B 5 s at 20°C	145	9·2	0·7	4·3	7·0	M	25·1
PP	None	—	—	—	0·25	0·28	I	—
PP	Conc. H_2SO_4 1h at 70°C	262	9·7	0·4	2·6	1·1	I	—
PP	Acid B 5 s at 20°C	382	14·5	0·3	3·3	2·8	I	35·9
PP	Acid A 1 min at 20°C	283	16·2	0·4	4·6	4·7	I	22·7
PP	Acid A 6 h at 70°C	261	13·5	0·4	4·0	11·2	M	13·7

[a] Acid A 'normal' chromic–sulphuric acid ($K_2Cr_2O_7$:H_2SO_4 = 7:12:150 by weight); Acid B as Acid A but 1/100th concentration with respect to $K_2Cr_2O_7$.
[b] I, Apparent interfacial failure; M, material failure.
[c] For a homogeneous sample O1s:O2s $\approx$ 10 .
Reproduced from reference 48 with permission.

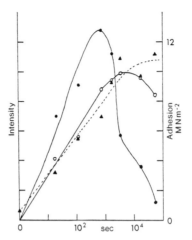

FIG. 8. CrO_3–H_2O etching of LDPE as a function of time at 25°C. ○, O1s peak intensity (10^4 counts s^{-1} fsd); ●, Cr2p$_{3/2}$ peak intensity (10^3 counts s^{-1} fsd); ▲, lap shear strength of adhesive joint with epoxide.[48]

the presence of structures such as

$$
\begin{array}{c}
\text{HO} \qquad \text{OH} \\
\diagdown \qquad \diagup \\
\overset{|}{\text{Cr(III)(H}_2\text{O)}_n} \\
| \\
\text{O} \\
| \\
-\text{CH}_2-\overset{\displaystyle |}{\underset{\displaystyle |}{\text{C}}}-\text{CH}_2- \\
\text{CH}_3
\end{array}
$$

The Cr is lost on oxidative chain scission or on acid hydrolysis.

Clearly the rate of oxidation will strongly depend on the concentration of the attacking species in solution. Figure 10, from the work of McGregor and Perrins,[47] illustrates the effectiveness of different compositions in the $CrO_3/H_3PO_4/H_2O$ system. There is a narrow region in the composition diagram which gives a much higher-than-average etching rate. No clear explanation was forthcoming, since IR spectra of an effective solution only detected H_3O^+ and HSO_4^- ions. These workers postulated, however, that the amount of free H_2O for combination with CrO_3 was important. This is consistent with H_2CrO_4 being the important species.

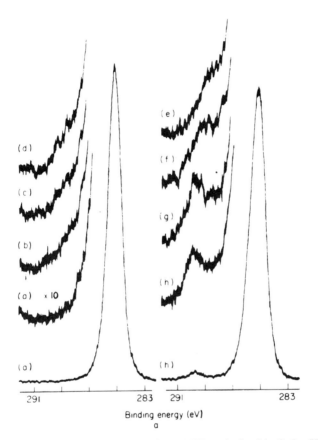

FIG. 9(a). High resolution C1s spectra for LDPE etched with CrO_3–H_2O at 25°C for the following times: (a) 0s, (b) 20s, (c) 40s, (d) 2 min, (e) 10 min, (f) 1 h, (g) 5 h, (h) 16 h. Count rate $= 3 \times 10^3$ counts s^{-1} fsd.[48]

5. EXTRUSION LAMINATION OF POLYETHYLENE WITH ALUMINIUM FOIL

The type of machinery used in this process is shown in Fig. 11. It is well established that, at typical extrusion temperatures (e.g. 280–300°C), LDPE will undergo rapid oxidation and therefore at the chill roll nip, aluminium foil contacts an oxidised polymer surface. Willis and Zichy[19] have correlated the degree of oxidation assessed by MIR–IR with PE–Al adhesion for a variety of extrusion temperatures and line speeds. However, in cases where end-use of the laminate requires the in-

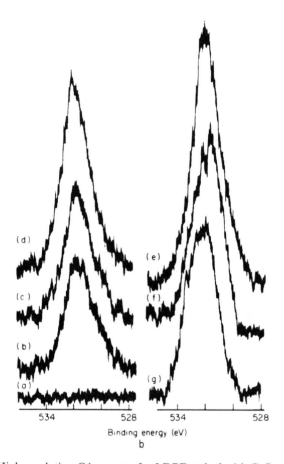

FIG. 9(b). High resolution O1s spectra for LDPE etched with CrO_3 –H_2O for the following times: (a) 0s, (b) 20s, (c) 40s, (d) 2min, (e) 10min, (f) 5h, (g) 16h. Note the spectrum for 1 h etching has been omitted for clarity; the intensity is $\sim 20\%$ greater than for peak (e) and the BE of the peak maximum is the same. Count rate $= 3 \times 10^2$ counts s^{-1} fsd,[48]

corporation of antioxidant into the PE, adhesion of polymer to foil can be dramatically reduced. The empirically developed solution to this problem involves the direction of an 'ozone shower' into the nip region (a form of surface pretreatment).

XPS studies[40] of the PE surface in contact with the Al foil (by dissolution of the latter in NaOH solution) from a number of laminates

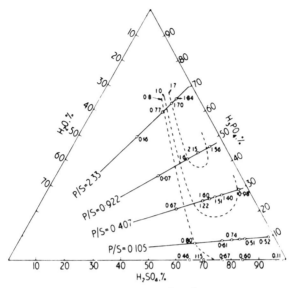

FIG. 10. Variation of etching rate (mg cm^{-2} h^{-1}) of polypropylene copolymer with composition of the etching acid (a mixture of CrO_3, H_2SO_4, H_3PO_4 and H_2O).[47]

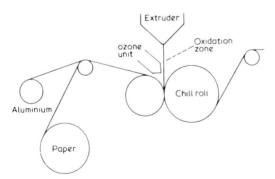

FIG. 11. Schematic diagram of process for lamination of aluminium with polyethylene.

made under different conditions are shown in Table 3. These data illustrate the dramatic effect of antioxidant, especially at the lower extrusion temperature. Note that XPS detects nitrogen-containing species besides oxidation products. Antioxidant has the obvious effect of retarding surface oxidation, thus reducing adhesion. The possibility that

TABLE 3
XPS DATA AND JOINT STRENGTHS FOR POLYETHYLENE EXTRUDED ON
TO ALUMINIUM

Extrusion conditions	O:C atomic ratio (%)	N:C atomic ratio (%)	Lap shear strength (MN m^{-2})	Failure[a] type
Not extruded. Pressed against PET film, 5 min at 150°C.	$\leq 0.6^b$	0·0	≤ 0.4	I
Extruded at 280°C.	7.0^b	3·0	5·1	I + M
Containing 0·02% antioxidant. Extruded at 280°C.	0.17^b	0·46	0·6	I
Containing 0·2% antioxidant. Extruded at 280°C.	0.0^b	0·0	0	I
Extruded at 300°C.	5·6	1·5	5·7	I + M
Extruded at 300°C with ozone shower.	6·8	1·9	6·3	I + M
Containing 0·02% antioxidant. Extruded at 300°C.	7·4	3·3	4·8	I + M
Containing 0·02% antioxidant. Extruded at 300°C with ozone shower.	8·4	4·2	5·7	I + M
Containing 0·2% antioxidant. Extruded at 300°C.	1·5	0·43	0	I
Containing 0·2% antioxidant. Extruded at 300°C with ozone shower.	6·2	2·1	3·3	I

[a] I = apparent interfacial failure between polyethylene and aluminium; M = failure of the polyethylene.
[b] Highest figure for film containing 0·2% antioxidant, some of which is probably detected on the surface.
Reproduced from reference 40 with permission.

the antioxidant acts as a weak boundary layer can clearly be dismissed. The most interesting conclusion to be drawn from this work, however, concerns the effect of the 'ozone shower'. In the absence of antioxidant the 'ozone shower' has little effect on the surface composition; when high levels of antioxidant are present the 'ozone shower' re-establishes a similar surface composition (in terms of both the oxygen and nitrogen concentrations). This suggests that the 'ozone shower' nullifies the suppressive effect of the antioxidant rather than bringing about surface oxidation by a different mechanism.

6. CONCLUSIONS

Despite the fact that the polyolefin pretreatments described in this chapter have been in use for many years, a detailed understanding of the mechanisms by which plastic surfaces are modified and the reasons for subsequent improvement in adhesive performance is yet to be attained. The impact of surface analytical techniques, particularly XPS, over the last few years has been significant, providing information on the nature of new functionalities incorporated into the surface and on their depth distribution. The emerging theme would appear to be that pretreatments improve wettability via general oxidation but the improved adhesion may also depend on specific interactions, such as hydrogen bonds, involving only a small proportion of these new functional groups.

REFERENCES

1. Brewis, D. M. and Briggs, D., *Polymer*, **22** (1981), 7.
2. Gray, J. in *Plastics:surface and finish*, S. H. Pinner and W. G. Simpson (Eds.), Butterworths, London 1971, p. 29.
3. Manley, T. C., *Trans. Electrochem. Soc.*, **84** (1943), 83.
4. British Patent 1253630 (1971).
5. Levitzky, J. J., Lindsey, F. J. and Kaghan, W. S., *SPE J.* (1964), 1305.
6. Carley, J. F. and Kitze, P. T., *Polym. Eng. Sci.*, **20** (1980), 330; Kitze, P. T., Ph.D. Thesis, University of Colorado (1973).
7. US Patent 2859480 (1958).
8. US Patent 4029876 (1977).
9. Krager, R. and Potente, H., *J. Adhes.*, **11** (1980), 113.
10. van der Linden, R., *Kunstoffe*, **69** (1979), 71.
11. Gregory, B. H. and Carroll, R. J., *Proc. 1976 Tappi Internat. Seminar on Adhesive Labelling*, p. 16.
12. Kim, C. Y., Evans, J. and Goring, D. A. I., *J. Appl. Polym. Sci.*, **15** (1971), 1365.
13. Evans, J. M., *J. Adhes.*, **5** (1973), 1.
14. Stradal, M. and Goring, D. A. I., *Can. J. Chem. Eng.*, **53** (1975), 427.
15. Owens, D. K., *J. Appl. Polym. Sci.*, **19** (1975), 265.
16. Blythe, A. R., Briggs, D., Kendall, C. R., Rance, D. G. and Zichy, V. J. I., *Polymer*, **19** (1978), 1273.
17. Briggs, D. and Kendall, C. R., *Polymer*, **20** (1979), 1053.
18. Chalmers, J. M., Kendall, C. R. and Briggs, D., unpublished results.
19. Willis, H. A. and Zichy, V. J. I. in *Polymer surfaces*, D. T. Clark and W. J. Feast (Eds.), Wiley, New York 1978, p. 287.
20. Briggs, D. and Kendall, C. R., *Adhesion and Adhesives* (1982), in press.
21. Mitchell, J. Jr., and Perkins, L. R., *Appl. Polym. Symp.*, **4** (1967), 167.
22. Courval, E. J., Gray, D. G. and Goring, D. A. I., *J. Polym. Sci., Polym. Lett. edn.*, **14** (1976), 231.
23. Konieczko, M. B., Ph.D. Thesis, Leicester Polytechnic (1979).
24. (a) H. H. G. Jellinek (Ed.), *Aspects of degradation and stabilisation of polymers*, Elsevier, Amsterdam 1978.

226   D. BRIGGS

(b) E. Geuskens (Ed.), *Degradation and stabilisation of polymers*, Applied Science, London 1975.
25. Evans, J. M., *J. Adhes.*, **5** (1973), 9.
26. Briggs, D. and Kendall, C. R., unpublished results.
27. Stradal, M. and Goring, D. A. I., *J. Adhes.*, **8** (1976), 57.
28. Sharples, L. K., *Plastics and Polymers* (April 1969), 135.
29. Clark, D. T. and Dilks, A., *J. Polym. Sci., Polym. Chem. edn.*, **17** (1979), 957.
30. Briggs, D., Brewis, D. M. and Konieczko, M. B., unpublished results.
31. Briggs, D., unpublished results.
32. Bartusch, W., *Verpackungsrundschau*, **27** (1976), 87.
33. Sweeting, O. J., *The science and technology of polymer films*, Vol. 2, Wiley, New York 1971.
34. Hollahan, J. R. and Bell, A. T. (Eds.), *Techniques and applications of plasma chemistry*, Wiley, New York 1974.
35. Hurst, C. W. and Schanzle, R. E., *Mod. Packag.*, **40**(2) (1966), 163.
36. Leeds, S., *TAPPI*, **44**(4) (1961), 244.
37. Ayres, R. L. and Shofner, D. L., *SPE J.*, **28**(12) (1972), 51.
38. Briggs, D., Brewis, D. M. and Konieczko, M. B., *J. Mat. Sci.*, **14** (1979), 1344.
39. Buchel, K., *Adhesion*, **12** (1966), 508.
40. Briggs, D., Brewis, D. M. and Konieczko, M. B., *Eur. Polym. J.*, **14** (1978), 1.
41. Gaydon, A. G., *The spectroscopy of flames*, 2nd edn., Chapman and Hall, London 1974.
42. Ghorashi, H. M., *Plating and Surface Finishing* (April 1977), 42.
43. Briggs, D., Brewis, D. M. and Konieczko, M. B., *J. Mat. Sci.*, **11** (1976), 1270.
44. Perrins, L. E. and Pettett, K., *Plastics and Polymers* (December 1971), 391.
45. Morris, C. E. M., *J. Appl. Polym. Sci.*, **14** (1970), 2171.
46. Blais, P., Carlsson, D. J., Csullog, G. W. and Wiles, D. M., *J. Coll. Interface Sci.*, **47** (1974), 636.
47. McGregor, A. and Perrins, L. E., *Plastics and Polymers* (June 1970), 192.
48. Briggs, D., Zichy, V. J. I., Brewis, D. M., Comyn, J., Dahm, R. H., Green, M. A. and Konieczko, M. B., *Surf. Interface Anal.*, **2** (1980), 107.
49. Isaacs, N. S., *Reactive intermediates in organic chemistry*, Wiley, New York 1974, p. 445.

Chapter 10

SURFACE TREATMENTS FOR POLYTETRAFLUOROETHYLENE

R. H. Dahm

Leicester Polytechnic, Leicester UK

1. INTRODUCTION

Polytetrafluoroethylene (PTFE) is a material notable for its high thermal stability and high resistance to chemical attack,[1,2] as well as for its unique surface properties such as its extremely low surface energy[3] and unique antifrictional properties.[4,5]

In order to utilise fully these unique properties, it is frequently necessary to bond PTFE adhesively, a process that is virtually impossible without first subjecting the polymer surface to one of a number of pretreatments. Quite apart from its technological importance, PTFE has been the subject of a large number of fundamental adhesion studies, and some of the more important results obtained in this area since the original papers by Benderly[6,7] are included in this review.

Considerable progress has been made in elucidating the mechanism of the reductive treatments and a major part of this review is concerned with the progress made in this field.

PTFE is generally regarded as one of the least chemically reactive of the polymers and the surface of this material must be activated by one of the methods described below in order to promote adhesion. The following brief survey is not intended to be complete, but serves to illustrate the large number of techniques that have been used.

(a) Immersion in a solution of an alkali metal in liquid ammonia.[8]

(b) Immersion in a solution of an alkali metal radical anion salt, typically sodium naphthalenide, in a polar aprotic solvent such as tetrahydrofuran.[6,9]

(c) Reduction with an electrochemically generated tetra-alkyl-ammonium radical anion salt solution.[10]

(d) Treatment at elevated temperatures with alkali metal vapours, a ternary sodium/tin/lead alloy, or sodium or calcium hydride.[11]

(e) Treatment with alkali metal amalgams at ambient temperature or above.[12,13]

(f) Direct electrochemical reduction of the surface when placed in contact with a metal cathode in a non-aqueous electrolyte at ambient temperature.[14]

(g) Exposure to molten potassium acetate at 325°C.[15]

(h) Exposure to a corona discharge in an atmosphere of hydrogen or dry ammonia.[16]

(i) Irradiation with an electron beam (to improve the adhesion of gold to PTFE).[17]

(j) Metal sputtering of the PTFE surface.[18]

(k) Vacuum deposition of metals on to the polymer surface with or without subsequent removal.[19-21]

(l) Exposure of the polymer to an inert gas such as neon or helium in a glow discharge, the so-called CASING procedure.[22,23]

(m) Modification of the surface by plasma or radiation-induced grafting.[24]

Several of the pretreatments listed above involve reduction of the PTFE surface to form a coherent, dark, carbonacious film which can be readily bonded using, for example, epoxides. However, only the two first-mentioned techniques have found wide commercial acceptance since the treatment is both rapid and inexpensive. Goldie[25] gives details for the preparation of the treatment solutions, and solutions of sodium naphthalenide are available commercially, e.g. under the trade name 'Tetraetch' (W. L. Gore Associates Inc.). Although few of the other treatments are of commercial importance at present, much insight has been gained into the mechanism of the reductive treatments from studies of the amalgam reductions and direct electrochemical reductions, while the studies of metal deposition, CASING, etc., have contributed greatly to the fundamental understanding of the bonding process in the adhesion to polymers, a topic that has been the subject of a number of recent reviews.[26-30]

2. MECHANISMS OF PRETREATMENTS

2.1. Sodium–liquid ammonia and sodium naphthalenide treatments

Immersion of PTFE (and a number of other fluoropolymers) for a few seconds in one of the above solutions containing about 15–23 g of sodium per litre results in the formation of a brown-black, strongly adherent, surface layer of thickness 0·05–1 μm.[31] The treatment results in a very large increase in bond strength, values in excess of 11 MNm^{-2} (1600 psi) being readily attainable in single lap shear tests.[7] The treatment involves reduction of the polymer surface to yield a carbonaceous material, the precise nature of which has been the subject of a number of studies. Borisova et al.,[15] using infrared spectroscopy of a composite made up of seven 5 μm-thick foils, showed that the alkali metal and the molten potassium acetate treatments result in blackened surfaces containing OH, CO, C—C, C=C, CH$_3$ and CH$_2$ species as well as NH groups for samples treated with sodium in liquid ammonia. The thickness of the treated layer was estimated to be of the order of tens of microns, and the OH,CO and hydrocarbon species were attributed to secondary processes taking place on the surface.

Brecht, Mayer and Binder,[32] Dwight and Riggs,[33] and Dwight[31] investigated the treated surface independently using ESCA. Both groups showed that the defluorinated surface consists mainly of carbon and oxygen. Brecht showed that the oxygen is present as C=O, COH and CO$_2$H produced after exposure of the treated surface to air, since no oxygen was found if the film was handled in an inert atmosphere, in which case the surface was found to consist almost entirely of carbon. Dwight further showed that the surface is highly polar, as shown by a large decrease in contact angle, and that the carbonaceous material can be removed completely by oxidising agents such as aqueous sodium hypochlorite solution or air in the presence of ultraviolet light, to regenerate the original polymer surface. PTFE is therefore unique amongst polymers in that the treated surface layer consists of a new, more-or-less homogenous phase which can be completely removed by chemical means. The reductive treatment of PTFE therefore involves much more than the mere modification of the surface but rather the *complete* conversion of a definite and quite thick layer of PTFE into a form of carbon, the composition of which appears to remain constant up to the polymer–carbon interface.

2.2 Treatment with amalgams

Although the surface properties of treated PTFE have received considerable attention (see for example Schonhorn's review[26]), virtually no work dealing with the properties of the carbonaceous material, let alone the mechanism by which it is formed, has appeared. However, the reduction of PTFE by alkali metal amalgams in the temperature range 25–100°C has been extensively investigated by Jansta and Dousek, both with respect to the nature of the reduction product[34-36] and with respect to the mechanism of the reduction.[12,13,37,38] Thus, pieces of skived PTFE tape, 0·5 mm thick, when held beneath the surface of lithium amalgam (typically 1 at.% Li) in a sealed evacuated ampoule, turn completely black in approximately 30 minutes. The thickness of the black layer increases to about 100 μm in 56 days, resulting in the formation of a dull black, firmly adherent, brittle, electrically conducting layer. By determining the amount of lithium consumed, it was shown that the overall reduction can be represented by the following equation:

$$+ CF_2 +_n + 2nMHg_x \longrightarrow nC + 2nMF + 2xnHg$$

Although the reaction penetrates into the bulk of the polymer, no mercury was found within the carbonaceous layer, which was homogeneous and of uniform thickness over the whole surface of the foil, and consists of an intimate mixture of carbon and lithium fluoride. The following solid state corrosion mechanism was suggested for the penetration reaction. At the amalgam–PTFE interface, direct reduction leads to formation of a thin film, consisting of electronically conducting carbon and ionically conducting lithium fluoride. Thus electrons released by the amalgam anode are transported by the carbon to the PTFE–carbon interface, where more of the polymer is reduced to carbon and fluoride ion. The return path for this corrosion cell is provided by lithium fluoride, a known solid state Li^+ ion conductor.

Thus, at the carbon–amalgam interface, the following anodic reaction takes place

$$LiHg_x \longrightarrow Li^+ xHg + e$$

and at the carbon–PTFE interface

$$-CF_2- + 2e \longrightarrow -\dot{C}- + 2F^{\ominus}$$

followed by:

$$Li^+ + F^\ominus \longrightarrow LiF$$

with Li^+ being the mobile species. The overall rate of the penetration reaction was assumed to be controlled by the sum of the electronic and ionic conductivities, and the thickness of the treated layer was found to increase with the square root of the time.

2.3. Electrochemical treatment

The essential electrochemical nature of the reduction of PTFE was strikingly demonstrated by Brewis, Barker, Dahm and Hoy.[39] These workers showed that the surface of PTFE, when held in contact with a cathode at a potential more cathodic than -1.5 V versus the saturated calamel electrode (SCE) in an aprotic solvent containing a tetra-alkylammonium support electrolyte, is reduced to yield a black adherent film, growing outward from the point of contact. The appearance of the treated surface resembles that of chemically treated PTFE, exhibiting similar bonding and wetting characteristics. The growth is not uniform from the point of contact, but is more pronounced in the direction of skiving or extrusion for ram extruded rods.[40] The area of reduced material obtained in this way depends on the water content of the electrolyte, typically dimethylformamide (DMF) dried over molecular sieves and distilled, and on the processing history of the PTFE specimen. For sintered material, an area of $1-2$ cm^2 is treated in about $15-20$ s at room temperature, whereas the area of unsintered material is very much larger and is generally limited by the physical dimensions of the specimen. Anisotropic etching of PTFE has not been observed for chemically treated material and is probably due to the formation of a highly oriented atypical surface layer produced during processing. This is supported by the observation that specimens that have not been subjected to such a severe processing regime, such as pressed sheets or samples from which the atypical surface layer has been removed chemically, show essentially isotropic etching. The growth of the film appears to take place in two stages. In the initial stage, the PTFE surface is reduced relatively quickly in the form of thin 'dendrites', radiating outwards from the cathode. In the second stage, the material between the dendrites is reduced[14] to yield an almost circular reduced region on the PTFE surface. The mechanism of the reduction may be represented by the following scheme, involving reduction of PTFE at the electrode–

electrolyte–polymer interface by a mechanism which may be similar to that for the reduction of vicinal dihalides:[41]

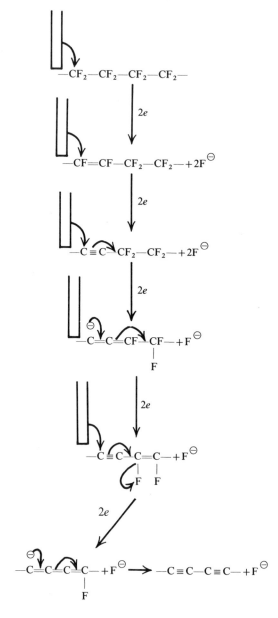

This leads to the formation of a polyacetylide by unzipping individual polymer chains. Formation of acetylenes by the electrochemical reduction of polyhalides is well established,[42] and polyacetylides produced by oxidation of copper acetylide are electronic conductors with a resistivity $\simeq 100\,\Omega m$.[43,44] The above scheme undoubtedly amounts to a severe over-simplification of the reduction mechanism, and the structure of the reduction product is probably complex. However, if it is assumed that the reduction product, whatever it is, is electronically conducting, then it is easy to see how the reduction can spread from the cathode over the remaining PTFE surface, since a new three-phase interface is continuously maintained. A considerable body of evidence in support of this mechanism has been assembled.[14] For example, application of a potentiostatic pulse to a point contact cathode results in a rising current–time transient exhibiting a maximum. This maximum is due to the potential drop across the film, which ultimately results in the lowering of the potential at its periphery to a value below the minimum required (-1.5 V SCE) for the reduction of PTFE. The slope of the rising part of the transient is a measure of the rate of the reduction, and its value depends on the nature of the PTFE surface as well as on the applied potential, showing that the reaction is under electrochemical rather than mass transport control.[14] Using values of the initial slope, obtained on ram extruded rod from which successive surface layers had been removed, the authors were able to obtain a value of $4\,\mu m$ for the thickness of the orientated atypical surface layer present on a ram extruded rod.[14] It was also possible to induce anisotropic etching by stretching dumbbell sections. PTFE fibres containing a very high concentration of aligned chains within the surface are reduced more rapidly than all other specimens.[45]

The rate of penetration of the electrochemical reduction into the bulk of the polymer has been determined and compared with the results obtained by Jansta and Dousek. The log current potential curves exhibit a linear Tafel region of ca 160 mV per decade, and a limiting current at potentials greater than 2.1 V SCE, the value of which decreases with increasing thickness of the carbonaceous film.[39] This strongly suggests that the rate of reduction is controlled by diffusion of some species through the carbonaceous film. This should give rise to a parabolic dependence of thickness on time for potentials on the diffusion plateaux, which is borne out in practice as shown in Figs. 8 and 9 of reference 14. The latter includes for comparison the data obtained by Jansta and Dousek[13,38] for alkali metal amalgams. The similarity in the rates for the two processes is striking, but it is unlikely that the electrochemical reduction proceeds by Jansta and Dousek's solid state mechanism, since tetra-alkylammonium fluorides are soluble in DMF and in any event are unlikely to be cationic conductors. It is also unlikely that the carbonaceous film could accommodate two of the large tetrabutylammonium cations for every carbon atom present. It is probable that the reaction involves diffusion of tetra-alkylammonium ions through the porous carbon film to the PTFE–carbon interface. The rate for such a process would be highly dependent on the size of the cation and, as shown by Brewis et al.,[39] the rate of reduction using the large tetraoctylammonium ion is nearly two orders of magnitude lower than that for the smaller tetraethylammonium ion.

It thus appears that the amalgam and direct electrochemical reductions proceed by similar mechanisms in which electrons are supplied to the PTFE surface by the electronically conducting carbonaceous reduction product, and the electrochemical return path is provided by a solid state lithium ion conductor in the one case, or by diffusion of tetra-alkylammonium ions through the solution-filled pores in the other. It is of interest to speculate which if any of these models applies to the chemical pretreatments. It is tempting to apply the Jansta and Dousek model to these systems since the alkali metal fluorides are virtually insoluble in both liquid ammonia and THF.[46] However, the limited amount of comparative data available do not entirely support the view (see Table 1).

TABLE 1

THICKNESS OF PTFE REDUCED BY CHEMICAL METHODS[49]

Treatment	Treatment time (h)	Thickness of PTFE reduced (μm) at 25°C [value for corresponding amalgam calculated from refs. 13 and 38]
Li naphthalenide/THF	240	9[44]
Na naphthalenide/THF	240	1[2·5]
K naphthalenide/THF	240	0·5
Na liquid NH$_3$	2·5(-50°C)	41[0·3]

Thus radical anion reductions appear to proceed at a slower rate compared with the corresponding amalgam, whereas sodium in liquid ammonia reacts very much faster than sodium amalgam. It is possible that, in these cases, PTFE is reduced directly by solvated electrons or radical anions diffusing through the solution-filled pores of the carbonaceous film. A similar mechanism involving amalgam-filled pores was ruled out by Jansta and Dousek owing to the high surface tension and wetting angle of the amalgam, although these workers do not appear to have considered the fact that the surface tension of mercury approaches zero at the highly cathodic potentials required for PTFE reduction. The stoichiometry of the reduction of PTFE to carbon requires a minimum of four electrons per C_2F_4 repeat unit. Direct measurement, however, shows that in the case of the amalgam[37] and electrochemical reductions,[14] a quantity slightly in excess of four electrons is obtained. In the case of the electrochemically produced carbonaceous material this excess charge amounts to one electron per 23 carbon atoms, and it has been suggested that it is due to the formation of a tetra-alkylammonium intercalation compound.[14] The formation of graphitic tetra-alkyl-ammonium compounds is well established.[47,48] However, as mentioned earlier and discussed in more detail by Barker,[49] the electrochemical reduction product is probably best represented by polyacetylide chains more or less cross-linked at random by sp^2 and sp^3 hybrid bonds. Such material should be capable of forming compounds similar to the graphite intercalation compounds, in which the excess electrons are held in the extended conjugated carbon chains and overall charge neutrality is maintained by an equivalent number of tetra-alkylammonium ions. Similar sodium-doped polyacetylenes have been described recently by Chaing et al.[50]

Barker et al. also determined the electronic conductivity of the carbonaceous material. The value varies from ca 10^{-4} Ωm for a freshly intercalated film to ca 10^2 Ωm for an anodically disintercalated film, and decreases rapidly and irreversibly to 10^4 Ωm upon exposure to air or protic solvents.[14] Other forms of carbon, for example graphite, are known to react rapidly and irreversibly with oxygen.[51] Examination of the carbon by ESCA always revealed the presence of oxygen, even when using glove box techniques.[14] Similar results were obtained by Dwight[31] and by Brecht et al.[32] for chemically grown films, although the reactivity to oxygen in this case appears to be considerably lower.

2.4. Nature of carbonaceous layer

Jansta, Dousek and co-workers have carried out a comprehensive investigation into the nature of the carbonaceous product obtained form the amalgam reduction of PTFE. They find that the product consists of a ceramic-like non-porous gas-impermeable mixture of carbon and lithium fluoride with an additional quantity of lithium (ca 2%w/w) bound to the

reduced polymeric carbon.[35,36] Using wide angle X-ray scattering, they
showed that the LiF crystallites are only about $3\,\mu m$ in size and the
carbon appears to be completely disordered, consisting of polymeric
chains containing (C=C) and (C≡C) groups, immobilised and separated from each other by salt sheaths. By determining the influence of
compression on the apparent density of the carbon, Jansta and Dousek
showed that a certain amount of cross-linking takes place after releasing
the carbon from its salt sheath by either melting out or dissolving out the
LiF. Dissolution of the LiF by water at room temperature results in the
spontaneous disintegration of the reduced film to form thin carbon
lamellae, which reflect the structure of the original primary PTFE
particles.[36,37]

The surface area of this form of carbon was found to be extremely high
at $4000\,m^2 g^{-1}$ and on contact with air at room temperature up to
$20\,wt\%$ of oxygen was irreversibly taken up by this material. It is
interesting to note that the surface area of electrochemically reduced
carbon is similarly in excess of $1000\,m^2 g^{-1}$ when determined from
double layer charging experiments.[39,52]

It thus appears that the reductive pretreatments of PTFE result in the
formation of a new phase consisting of a polymeric, non-graphitic form
of carbon which rapidly reacts with atmospheric oxygen, up to $20\,wt\%$
being incorporated into the film. Prolonged exposure to oxygen, particularly at elevated temperatures or in the presence of ultraviolet light, or
treatment with strong oxidising agents, results in the complete removal
of this film to give a surface displaying the same non-wetting and poor
adhesion properties as the original PTFE.[33,53] The reasons for this
behaviour are not clear. There is no measurable increase in the oxygen
content (as determined by ESCA),[33] nor is there evidence for residual
unsaturation.[49] This leaves dimerisation or cross-linking of free radical
chain ends as possibilities. This, according to the weak boundary layer
theory, might be expected to result in an increase in adhesive bond
strength which is certainly not the case in practice. Yet, in the case of
skived tapes treated with sodium naphthalenide, there are subtle differences between regenerated and 'virgin' PTFE surfaces.

2.5. Deactivation of PTFE towards sodium naphthalenide
The commonly used sodium naphthalenide etchant can only be used
reliably on PTFE exhibiting the atypical surface layer produced by
skiving or ram extrusion. Once this layer has been removed by repeated
reduction followed by oxidation of the carbonaceous film, further re-

duction results in the formation of a very thin, non-coherent carbonaceous surface which cannot be bonded successfully. Reactivity towards naphthalenide may be restored, however, either by heating the sample to approximately 50°C followed by rapidly quenching to 10°C, or by etching the surface electrochemically or with sodium in liquid ammonia, followed by the usual oxidative removal of the carbonaceous material. A tentative explanation in terms of the preferential reduction of crystalline material has been offered,[53] although other workers suggest that, in common with the polymeric hydrocarbons, the amorphous regions are in fact preferentially reduced.[54] The basic assumption is made that the ease of reduction of PTFE (as determined for example by its reduction potential) is not uniform along a given PTFE chain, but is somewhat greater at certain activated sites within the crystalline regions of the polymer. Thus, a relatively mild reducing agent such as sodium naphthalenide will attack preferentially at these sites, which act as nuclei for further reduction of the surface. Once all of these nuclei have been consumed (typically after 5–6 treatments for 0·25 mm skived tapes), further reduction becomes impossible. However, the heating and quenching cycle may result in the formation of new active sites formed, for example, by the rotational changes occurring when the polymer passes through the 19° or 30° transitions.[55] After approximately 25 treatment cycles, all reactive material will have been removed from the surface region and thermal activation does not occur. In this state the sodium naphthalenide treatment becomes totally ineffective. It is, however, possible to restore the reactivity of the surface by a treatment with the far more powerful reducing agent sodium in liquid ammonia, essentially a solution of solvated electrons, which is a sufficiently powerful reducing agent to react with both crystalline and amorphous material and hence expose fresh active material. Similar results may be obtained by physically abrading the deactivated surface or by subjecting it to an electrochemical treatment followed by the removal of the carbonaceous film.

3. SURFACE MODIFICATION OF PTFE PRODUCED BY CONTACT WITH SOLID METAL SURFACES

The various surface pretreatments discussed so far all proceed by a mechanism in which electrons are transferred to the polymer. The standard reduction potential calculated for this process from standard thermochemical data amounts to $+0.917\,V$ versùs NHE*or$+0.67\,V$ versus SCE. PTFE is thus thermodynamically unstable when in contact with most common metals, one of the few stable metals being gold. However, in practice, no reduction can be detected at potentials below $-1.5\,V$, so that the reduction requires a very large overpotential which is almost certainly kinetic in origin. This, together with the presence of oxide layers and the low conductivity of most metal fluorides, is largely responsible for the observed stability of this material, although care should be taken when PTFE is employed in electrochemical devices or in bearings where passive layers are continuously removed and high temperatures may be produced. A number of papers have appeared dealing with the effect of metals on perfluoropolymers. These fall into three major groups, described in sections 3.1–3.3.

3.1. Studies of morphological changes caused by melting fluoropolymers against metals

Studies have been carried out in which the molten polymer is allowed to crystallise in contact with a high energy surface such as gold or aluminium. Schonhorn and Ryan[56] showed that the tensile shear strength of joints made with PTFE film heated to 400°C in contact with an evaporated gold film followed by rapid cooling increased from *ca* $1.4\,MN\,m^{-2}$ for untreated skived tape to *ca* $10.3\,MN\,m^{-2}$; this value compares with *ca* $16.5\,MN\,m^{-2}$ obtained with sodium naphthalenide etched material. This treatment also resulted in a marked increase in the wettability of the surface, with an increase in the critical surface tension of wetting from *ca* $18\,mJ\,m^{-1}$ to $40\,mJ\,m^{-1}$. Even larger increases in bond strength were obtained when FEP was crystallised against gold, but evaporated aluminium was found to be much less effective.[57] According to Schonhorn, the above treatments result in the elimination of a weak boundary layer and its replacement by a mechanically strong surface layer with a high degree of crystallinity which, in contrast to the 'strong' surface produced by CASING, has a high surface free energy. Apparently, gold is far more effective in producing a mechanically strong surface region than is aluminium oxide and it has been suggested that this is due to the greater density of suitable nucleating sites on the gold surface.[30] An entirely different interpretation is provided by Dwight and Riggs, who found that the surface of the polymer had been changed chemically to yield a thin (a few tens of Ångstroms) surface layer

---

* Normal hydrogen electrode.

characterised by ESCA and shown to resemble the spectra obtained for sodium in liquid ammonia treated FEP.[33] Thus, although gold is thermodynamically stable in contact with PTFE at room temperature, it is quite possible that, at the high temperatures involved in these crystallisation experiments, the reduction of PTFE by gold is thermodynamically feasible. Unfortunately, no data are available for this system, although the reaction of PTFE and a number of other metals at temperatures in excess of 700 K has been investigated.[21]

3.2. Studies of chemical changes induced in fluoropolymer surfaces when contacted with metal surfaces under tribological conditions
Such studies are of interest because the effect of passive oxide and fluoride films is greatly reduced, since these are likely to be physically broken down under tribological conditions. Thus, Cadman and Gossedge[21] have investigated PTFE–metal interactions by heating mixtures of various powdered metals and polymer in a differential scanning calorimeter followed by ESCA examination of the reaction products. Some of these results are presented in Table 2.

TABLE 2
REACTIONS BETWEEN METAL AND PTFE HEATED IN THE DSC[21]

Metal	M.pt if <773 K	Comment
In	423	Black carbonaceous residue on heating to 733 K.
Sn	505	Black carbonaceous residue on heating to 733 K.
Cd	593	Black carbonaceous residue on heating to 683 K.
Zn	693	No reaction up to 703 K.
Al		No reaction up to 703 K.
Mg		No reaction up to 703 K.
Fe		No reaction up to 703 K.
Ni		No reaction up to 703 K.
Cr		No reaction up to 703 K.
Cu		No reaction up to 703 K.

These results clearly show that PTFE is reduced by a number of low melting metals (zinc is an important exception), whereas it does not react with solid metals, even the strongly electropositive ones, aluminium and magnesium. The reason for this behaviour is presumably due to the presence of passive oxide layers, in the case of the solid metals, preventing the electron transfer reactions from taking place. However, examination of aluminium, [21] stainless steel or nickel,[58] after rubbing with PTFE, showed the emergence of peaks in the ESCA spectrum which could be assigned to fluoride ion formed at the metal–polymer interface during the initial stages of the rubbing process.

3.3. Studies of surface changes caused by vacuum deposition of metals on to clean fluorocarbon surfaces

This is one obvious way of bringing about intimate contact between the polymer surface and the oxide-free metal, especially if the process is carried out in an ultra-high vacuum. Thus Roberts et al.[19] found that vacuum deposition of a layer of aluminium 100 nm thick on to FEP followed by its removal with dilute aqueous alkali greatly improves the adhesion of gold subsequently vacuum deposited on to the treated surface. Indeed, the gold to FEP joint strength now approaches the bulk strength of the polymer, whereas direct evaporation of gold on to the untreated polymer produces joints of virtually zero strength.

Examination of the treated surface by ESCA revealed the familiar pattern, i.e. a defluorinated surface region with an intense oxygen peak, the spectrum resembling that obtained by Dwight and Riggs[33] for FEP etched with sodium in liquid ammonia. The treated region is, however, very much thinner: 5–10 nm, compared with the 1 μm-thick films produced by the sodium naphthalenide treatment.[6]

Presumably, in this case, a corrosion mechanism similar to that proposed by Jansta and Dousek for the amalgam reduction operates with the highly electropositive aluminium acting as electron donor and some ionic aluminium species, lost from the surface when the aluminium was removed with sodium hydroxide solution, providing the return path of the corrosion cell. Indeed, Schonhorn and Roberts found evidence for the existence of interfacial organometallic species by examining the polymer surface by ESCA through thin films (1·0–6·0 nm) of aluminium or titanium, vapour deposited on to the FEP surface.[59] They also measured the adhesion between these vapour deposited films, about 10 nm thick, and the FEP substrate, using an epoxide in a conventional lap shear configuration. Their results are shown in Table 3.

The results clearly show that titanium is even more effective than aluminium in defluorinating the surface, and also forms the strongest bonds to that surface. Schonhorn and Roberts suggest a mechanism in

TABLE 3

ADHESION OF EVAPORATED METAL FILMS ON FEP[19]

	Tensile shear strength $(kg\ cm^{-2})$	Intensity of F_{1_s} / Intensity of C_{1_s}
FEP, no metallisation	~ 0	7·5
FEP, 6·0 nm Au vapour deposited	~ 0	5·7
FEP, 6·0 nm A1 vapour deposited	~ 94	1·6
FEP, 6·0 nm Ti vapour deposited	$\sim 142\cdot 5$	0·5

which the polymer is dehalogenated to yield an unsaturated species, which then reacts further to form a highly cross-linked surface in a reaction that is catalysed by the metal fluoride according to the following overall scheme:[60]

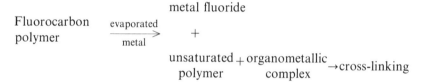

No evidence is presented for the formation of such a cross-linked surface in the case of fluoropolymers, and it is difficult to see how such a process could operate in the case of the sodium naphthalenide treatment, where the only possible catalyst is sodium fluoride. Vogel and Schonhorn[20] investigated the adhesion of a number of metal vapours deposited on to PTFE. Adhesion levels were considerably smaller than those obtained for FEP, and no significant change in wettability of PTFE (or poly-ethylene) due to the deposition, followed by subsequent removal, of metals such as titanium and aluminium, could be detected. However, Briggs and Gribbin,[60] using ESCA, found peaks characteristic of de-fluorinated carbon when examining PTFE surfaces after removal of vacuum deposited aluminium. Cadman and Gossedge have found peaks characteristic of fluoride ion when tin or indium was evaporated on to a clean PTFE surface in a UHV preparation chamber directly attached to the photoelectron spectrometer.[21] Significantly, no fluoride ion could be detected when gold was deposited under the same conditions. Cadmium and nickel were also found to produce peaks characteristic of fluoride ion, whereas no such peaks were found after evaporation of silver. It is possible that PTFE is much less reactive towards metals than FEP, producing a much thinner layer of reaction products, which are readily removed by dissolution after removal of the metal film. It may also be of interest to mention here the important observation made by Gossedge and Cadman that, if a carbon-rich PTFE surface produced by argon ion etching is allowed to stand in the spectrometer, PTFE molecules gradually reappear in the surface region over a time period of several hours. Cadman and Gossedge[21] suggest that this is due to diffusion of mobile, possibly low molecular weight, PTFE into the surface region, through the layer of reaction products. Clearly, further work in support of this interesting observation is desirable.

It thus appears that the mechanism by which metals, vapour deposited on to PTFE, adhere to the substrate, is still not settled. Vogel and Schonhorn's[20] view appears to be that the polymer surface reacts with the metal to yield a cross-linked, mechanically strong surface, which exhibits the same wetting properties as the original polymer, so that the increase in adhesion is mainly due to the elimination of a weak boundary layer, rather than due to an increase in the surface free energy brought about, for example, by the introduction of polar functional groups into the surface. However, examination of these surfaces by ESCA seems to suggest quite severe reduction, leading to almost complete defluorination, so that the treated surface is probably better described as a separate phase, consisting of cross-linked polymeric carbon in analogy with the surface produced by treatment with alkali metal solutions. It is likely that a conventionally cross-linked PTFE surface would retain the low chemical reactivity of this material, whereas experiments clearly show that oxidation of the treated surface results in the regeneration of virgin PTFE.[33]

4. SURFACE MODIFICATION OF PTFE AND OTHER FLUORINE-CONTAINING POLYMERS BY GLOW DISCHARGES

Treatment of polyolefin films by means of a corona discharge is a well-established, commercially used technique that results in large improvements in the bondability and printability of these materials (Chapter 9 and reference 28). Attempts have been made to apply this technique to fluoropolymers but the technique does not appear to have gained wide acceptance. One method[16] involves exposure of the film to a corona discharge in an atmosphere of hydrogen or ammonia. The treatment yields bond strengths comparable to those obtained using the dissolved alkali metals. In another treatment,[61] the polymer is exposed to a discharge in a nitrogen atmosphere containing a small amount of an organic compound such as glycidyl methacrylate. The resulting film can be bonded readily and peel strengths up to $157 \cdot 5 \, \text{kg m}^{-1}$ can be achieved. Schonhorn and Hansen[22] found that exposure of polyolefins or perfluoropolymers to excited species, produced in a low power radio frequency electrodeless discharge in an inert gas such as helium, results in a large improvement in bonding, giving joint strengths comparable to those resulting after chemical etching, without affecting either the bulk properties such as colour and tensile strength or the surface properties such as wettability and surface conductivity. Schonhorn, using ATR, found evidence for unsaturation which could be readily eliminated by addition of bromine. The fact that treatment with bromine did not affect the bond strength of discharge treated PTFE was interpreted as evidence for the formation of a cross-linked surface layer of high cohesive strength, which is chiefly responsible for the increase in bond strength. Further evidence in support of this theory comes from the observation that perfluorokerosine can be converted from a volatile liquid to a hard polymeric solid of much higher molar mass when subjected to the above discharge treatment. Schonhorn coined the acronym CASING (Crosslinking by Activated Species of Inert Gases) for this surface treatment, and this term has become widely accepted.

Although no progress appears to have been made in an understanding of the mechanism of CASING when applied to perfluoropolymers, the application of ESCA to the particular case of ethylene and ethylene–tetrafluoroethylene copolymers has yielded much insight into the mechanism of the gas discharge treatment. Thus, in an early review, Clark and Feast[62] showed that argon ion bombardment of the ethylene-tetrafluoroethylene copolymer results in the gradual loss of fluorine from the outermost 5 nm of the sample, resulting in the conversion of $\underline{CF_2}$ sites to $\underline{CF}$ types. The following reaction scheme was suggested:

$$M^* + -CF_2-CF_2-CH_2-CH_2- \longrightarrow -CF_2-CF_2-\dot{C}H-CH_2- + H^\cdot + M$$

where M^* is an electronically excited species of argon.

The hydrogen atoms produced are now trapped in a fairly rigid matrix and may abstract hydrogen or fluorine atoms from neighbouring CH_2 or CF_2 groups. Combination of the radicals produced may then lead to the formation of cross links, i.e.

$$
\begin{aligned}
-CF_2-CF_2-CH_2-CH_2 + 2H^{\cdot} &\longrightarrow -CF_2CF_2-\dot{C}HCH_2- \\
&\quad\;\; +-CF_2\dot{C}F-CH_2CH_2- \left.\right\} \longrightarrow \\
&\quad\;\; +HF+H-H
\end{aligned}
$$

$$
\begin{aligned}
&-CF_2-CF_2-CH-CH_2 \\
&\qquad\qquad\qquad\quad | \\
&\qquad\qquad -CF_2-CF-CH_2-CH_2-, \text{ etc.}
\end{aligned}
$$

An analogous mechanism for perfluoropolymers, involving fluorine atoms as intermediates, was considered to be energetically less favourable.

In a subsequent publication,[63] Clark and Dilks determined the rate constants for the surface reactions and those taking place in the subsurface and bulk. The former were found to be an order of magnitude larger than the subsurface reactions, and it was concluded that the surface reaction is dominated by processes involving ionic and metastable argon species, whereas the reactions taking place within the subsurface and bulk are mainly brought about by radiative energy transfer. In all cases, the total integrated intensity of the C1s levels increases after ion bombardment, which is interpreted in terms of cross-linking processes brought about by the plasma in which CF_2 structural features are converted to CF environments. The binding energies of the latter were found to be consistent with structural features such as $-CF_2-CF-C-$ and $-CF-CF-C-$, in which CF groups are directly attached to CF_2 or CF groups, or to carbon atoms devoid of fluorine.

These results clearly show that exposure of ethylene–fluoroethylene copolymer to an argon plasma results in cross-linking of the polymer surface rather than in complete reduction to carbon as observed in the case of the chemical treatments of many perfluoropolymers. Clark and Dilks propose a reaction scheme[63] (Scheme 1).

The primary process is the formation of the parent ion of a portion of the polymer chain. This may then undergo disproportionation with elimination of a hydrogen atom (elimination of a carbon radical was considered unlikely owing to the cage effect) and formation of a carbocation.

A more important process involves capture of an electron by the parent ion to yield an electronically excited chain segment with internal energy sufficient to bring about elimination of small molecules, e.g. HF and H_2, to yield unsaturated centres, and elimination of atomic hydrogen

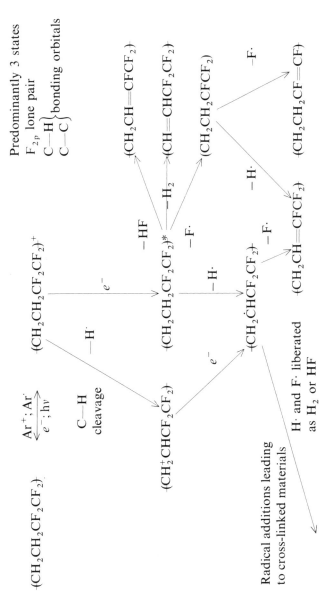

SCHEME 1.

or fluorine to yield carbon radicals in close proximity to the unsaturated centres, thus facilitating formation of cross links.

In a subsequent study,[64] Clark and Dilks found that helium is the most efficient gas for cross-linking the outermost few monolayers, whereas cross-linking of the subsurface and bulk polymer is best effected by a neon plasma.

The results obtained by the above workers thus lend support to Schonhorn's views regarding the formation of a mechanically strong cross-linked surface being a prime requirement for the formation of strong adhesive bonds to polymeric materials. However, other workers have noted that the resultant products from such experiments contain oxygen (see below).

Further evidence in support of cross-linking in PTFE is claimed by Smolinsky and Vasile,[65] who analysed, using mass spectrometry, the products obtained when freshly deposited plasma-polymerised PTFE is subjected to ion etching by noble gases (He, Ne, Ar and Xe) as well as reactive gases such as oxygen and hydrogen. The principal products obtained from noble gas discharge were C^+, CF^+, CF_2^+ and CF_3^+, with CF^+ being the major species. The fact that little etching occurred with hydrogen is surprising in a way, in view of the fact that the abstraction of fluorine by atomic hydrogen is thermodynamically feasible, although the products may of course be involatile. On the other hand, exposure to an oxygen-containing plasma produced the largest yield of ions, which, in addition to the types described above, also included the following oxygen-containing species: CO^+, CO_2^+ and COF^+. Fragments containing more than one carbon atom are rare whatever type of plasma is used. It thus appears that, in the case of PTFE, the mechanism involves a carbon–carbon bond breaking step and elimination of a single carbon-containing species, perhaps a difluorocarbene, rather than dimerisation of radicals formed by elimination of fluorine.

Yasuda, Marsh and Reilley[66] have examined surfaces of PTFE treated with an inert (argon) or reactive (nitrogen) plasma using ESCA. In both cases, considerable defluorination is observed, as evidenced by the decrease in the F1s and C1s peak intensities. They also find that oxygen is introduced for both the Ar and N_2 plasma, and nitrogen can be detected after treatment with the N_2 plasma, although the N1s and O1s peak intensities are only one half of those observed when polyethylene is subjected to a similar treatment. It thus appears that reactions other than simple cross-linking play an important role in the plasma treatment of PTFE. The introduction of polar groups containing oxygen should result in considerable changes in the wettability and critical surface tensions of the polymer surfaces. Such changes have indeed been observed by Hirotsu and Ohnishi[67] for a number of fluorinated polymers as well as for PE after exposure to He, N_2 and O_2 plasmas. As shown in Table 4, the fractional polar contribution to the total critical surface energy γ_S^p/γ_S (for a detailed discussion of critical surface energy concepts see Chapter 6 and reference 68), does indeed increase in all cases.

TABLE 4

FRACTIONAL POLAR CONTRIBUTION IN THE TOTAL CRITICAL SURFACE ENERGY OF UNTREATED AND PLASMA TREATED POLYMERS[67]

Treatment		γ_S^p/γ_S			
		PTFE	TFE–ET	PVdF	PE
(1) Untreated		0·00	0·05	0·14	0·08
(2) He	determined immediately after plasma treatment	0·08 $(2·0)^a$	0·12 $(1·7)^a$	0·26 $(8·3)^a$	0·34 $(7·3)^a$
N_2		0·07 (7·3)	0·20 (4·5)	0·26 (9·1)	0·36 (9·0)
O_2		0·15 (56)	0·25 (34)	0·26 (106)	0·42 (42)
(3) He	determined 3 days after plasma treatment	0·11	0·20	0·29	0·38
N_2		0·07	0·09	0·16	0·33
O_2		0·03	0·21	0·22	0·42

^{a} Numbers in parentheses refer to the degradation rate $\times 10^2$ mg cm^{-2} h^{-1} of the above polymers in the corresponding plasma.[67]
Abbreviations: PTFE, polytetrafluoroethylene; TFE–ET, tetrafluoroethylene–ethylene copolymer; PVdF, poly(vinylidene fluoride); PE, polyethylene.

These results suggest that plasma treatment results in the introduction of polar groups into the polymer surface. The effect is large in the case of PE when compared with PTFE. This is surprising, since the degradation rate of PTFE in an oxygen plasma is higher than that of other polymers, with the exception of PVdF. It thus appears that PTFE is highly sensitive to an oxygen plasma, in agreement with the mass spectroscopy results[65] referred to previously. However, PTFE is unique in that the plasma treatment results in degradation of the polymer without apparently introducing a large amount of surface functionality. Indeed, as shown in the following section, large differences in behaviour are frequently encountered when dealing with surface treatments involving

non- or partially fluorinated polymers on the one hand, and perfluorinated polymers on the other.

5. TREATMENTS INVOLVING THE CRAFTING OF A NEW POLYMERIC PHASE ON TO THE PTFE SURFACE

Such treatments are based on the generation of free radicals (or possibly
some other reactive intermediates) within the PTFE surface in the
presence of the polymerisable species. Thus, Mashonov and Avny[69]
coated a number of polymer films with glow discharge polymerised
acetylene and measured the tensile shear strength of the composite A1-
epoxy adhesive–acetylene plasma treated film-epoxy adhesive–A1. The
results obtained are summarised in Table 5.

TABLE 5
LAP SHEAR STRENGTHS FOR THE COMPO-
SITES A1–EPOXY ADHESIVE ACETYLENE
PLASMA TREATED FILM–EPOXY ADHESIVE–
A1[69]

Film thickness (mm)	Tensile shear strength ($MN\,m^{-2}$)	
	Untreated	Treated
PE (0·10)	0·62	8·62
PVF (0·03)	0·07	10·00
PVC (0·20)	6·21	10·00
PTFE (0·20)	0·07	1·72

The results presented in Table 5 clearly show that, although there is a
large percentage increase in bondability with PTFE, the resultant bond
strength is the lowest of the four polymers. An alternative procedure for
the activation of polymer surfaces followed by grafting of a suitable
monomer was investigated by Brenman and Lerchenthal.[70,71] The surfaces were activated by abrasion with emery paper in the presence of a
reactive adhesive or a primer such as methyl methacrylate. The authors
suggest that the lower reactivity of PTFE is due at least in part to the
creation of free radicals, rigidly held by the close-packed PTFE chains
and sterically quite inaccessible for reaction with the primer. However, as

shown in the following section, the surface of PTFE has now been successfully modified, using a radiation induced grafting technique that results in bond strengths superior to those obtained for sodium etched films when using peel tests, although lap shear bond strengths are inferior.

6. GENERAL DISCUSSION

Treatments with solutions of sodium naphthalenide or with solutions of sodium in liquid ammonia are still the most extensively used pretreatments for PTFE and some other fluoropolymers. The solutions are relatively inexpensive and the treatment, particulary with Na–liquid NH_3, is very fast, and may be used for the continuous treatment of PTFE tape. However, the above techniques suffer from the following disadvantages:

(a) The 'shelf life' of solutions of sodium naphthalenide in THF are fairly short, unless air and moisture are rigorously excluded. One recent patent[72] claims greatly improved shelf life by replacing THF by tetramethylurea, and Brockmann[29] recommends the use of a layer of petroleum ether to minimise ingress of air.

(b) Single sided treatment of skived tape is difficult but frequently desirable.

(c) Treated film does not store well, particularly when exposed to light,[33] which results in the gradual fading of the carbonaceous film. Attempts to regenerate the carbonaceous film by re-immersion in sodium naphthalenide are generally unsuccessful.[73]

(d) The joint strengths of sodium–etched PTFE film decrease upon exposure to ultraviolet light. In peel tests, failure was shown to occur at the carbonaceous film–PTFE interface and composites made up from PTFE tape containing carbon black possessed substantially better UV stability than composites prepared from unfilled PTFE. Further small increases in durability could be obtained by incorporating an antioxidant into the adhesive.[74]

It is unlikely that the above problems will be overcome by future improvements in the chemical etching procedures. On the other hand, considerable improvements are likely in the various plasma-induced treatments such as CASING and in the various grafting techniques, many of which have been successful for the pretreatment of polyolefins.

A notable success along these lines has been achieved by Yamakawa[75] for the radiation-induced grafting of methyl acrylate (MA) vapour on to PTFE, followed by saponification of the ester. The structure of the PTFE grafts were studied in great detail using ESCA, ATR and interference microscopy. The thickness of the grafted MA layer passes through a maximum after a dose of 1.5×10^5 rad. This dose is much lower than that required for extensive degradation of the fluoropolymer chains, so that no appreciable deterioration of the mechanical properties of the polymer occurs. This dose rate also corresponds to the maximum value in lap shear tests yielding values of up to $9.8 \, MNm^{-2}$ with failure of the PTFE adherend. As shown in Table 6, the increase in peel strength was even more dramatic, to give values much higher than those obtained for sodium-etched material.

TABLE 6

PEEL STRENGTHS OF SURFACE MODIFIED PTFE SHEETS BONDED WITH AN EPOXY ADHESIVE[75]

Surface treatment	PTFE sheet	Peel strength (kg/25 mm)
MA grafted and then saponified	Heat press formed	> 50
Na naphthalenide	Heat press formed, 2.0 mm thick	0.5–1.2
Na naphthalenide	Skived, 1.5 mm thick	4.2–5.5

Of equal importance is the fact that the homopolymer layer on PTFE resembles the PE grafts in being stable to abrasion, heat ageing and artificial weathering, so that the mass production of pretreated PTFE sheet is possible in principle.

The observation made by Galembeck et al.,[76] that low energy density liquids such as iron pentacarbonyl, $Fe(CO)_5$, are absorbed in significant amounts by PTFE, may offer an alternative approach to the surface pretreatment of fluoropolymers. Thus, Galembeck found that immersion of PTFE films in $Fe(CO)_5$ for one hour, followed by two hours in alkaline permanganate solution, results in the formation of an ultrafine dispersion of metal oxides, accompanied by a large increase in the wettability of the treated PTFE surface. However, Gribbin[77] found that this treatment did not result in a significant increase in bond-ability.

7. CONCLUSIONS

Surface pretreatment of PTFE may be effected by one of the following techniques:

(a) Exposure to a chemical or electrochemical reducing environment.
(b) Exposure to an inert or reactive gas plasma.
(c) Grafting of a new polymeric phase on to the PTFE surface.

The most widely used commercial treatments are those in group (a) and reduction with sodium napthalenide or with sodium in liquid ammonia are of particular importance.

The mechanism of these treatments involves complete reduction of the fluoropolymer to yield a mechanically strong, well-defined layer of carbonaceous material. This behaviour is unusual amongst polymers, in that treatment usually involves introduction of functional groups into the surface rather than conversion into a new phase. The treated fluoropolymer is unique, in that the carbonaceous layer can be completely removed chemically to reveal the original PTFE surface. The reactivity of the regenerated PTFE surface is reduced, in many cases, to such an extent that the widely used sodium naphthalenide treatment becomes ineffective unless special measures are taken to reactivate the surface.

Radiation-induced grafting of a new polymeric phase on to the PTFE surface offers a possible alternative treatment, if practical difficulties can be overcome. A major disadvantage of the reductive treatment arises from the high reactivity of the carbonaceous material towards oxygen, particularly in the presence of strong sunlight. This can lead to premature failure of joints and the alternative treatment described above may well yield considerable improvements in lightfastness.

REFERENCES

1. Koo, G., *Fluoropolymers*, L. A. Wall (Ed.), Wiley–Interscience, New York 1972.
2. Mark, H. F. and Gaylord, N. G. (Eds.), *Encyclopaedia of polymer science and technology*, Vol. 13, Wiley, New York 1970.
3. Zisman, W. A in *Contact angle, wettability and adhesion, ACS Adv. Chem. Ser.*, **43** (1964), 9.
4. Pascoe, M. W., *Tribology*, (1973), 184.

5. Richards, M. O. W. and Pascoe, M. W. in *Advances in polymer friction and wear*, Vol. 5A, L.-H. Lee (Ed.), Plenum Press, New York, 1974, p. 123.
6. Nelson, E. R., Kilduff, T. J. and Benderly, A. A., *Ind. Eng. Chem.*, **50** (1958), 329.
7. Benderly, A. A., J. *Appl. Polym. Sci.*, **6** (1962), 221.
8. Minnesota Mining and Manufacturing Co., Brit. Patent, 765, 284 (1957).
9. Rappaport, G., US Patent 2,809,130 (1957).
10. Brewis, D. M., Dahm. R. H. and Konieczko, M. B., *Makromol. Chem.*, **43** (1975), 191.
11. Purvis, R. J. and Beck, W. R., US Patent 2,789,063 (1957).
12. Jansta, J., Dousek, F. P. and Řiha, J., *J. Electroanal. Chem.*, **38** (1972), 445.
13. Jansta, J., Dousek, F. P. and Řiha, J., *J. Appl. Polym. Sci.*, **19** (1975), 3201.
14. Dahm, R. H., Barker, D. J., Brewis, D. M. and Hoy, L. R. J., in *Adhesion—4*, K. W. Allen (Ed.), Applied Science Publishers, London, 1979, p. 215.
15. Borisova, F. K., Galkin, G. A., Kiselev, A. V., Korelev, A. Y. and Lygin, V. I., *Kolloidn. Zh.* **27**(3) (1965), 320.
16. Ryan, D. L., *Brit. Patent*, 890,466 (1962).
17. Sessler, G. M., West, J. E., Ryan, F. W. and Schonhorn, H. J., *J. Appl. Polym. Sci.*, **17** (1973), 3199.
18. Mattox, D. M. and McDonald, J. E., *J. Appl. Phys.*, **34** (1963), 2493.
19. Roberts, R. F., Ryan, F. W., Schonhorn, H., Sessler, G. M. and West, J. E., *J. Appl. Polym. Sci.*, **20** (1976), 255.
20. Vogel, S. L. and Schonhorn, H., *J. Appl. Polym. Sci.*, **23** (1979), 495.
21. Cadman, P. and Gossedge, G. M., *J. Mat. Sci.*, **14** (1979), 2672.
22. Schonhorn, H. and Hansen, R. H., *J. Appl. Polym. Sci.*, **11** (1967), 1461.
23. Schonhorn, H., Ryan, F. W. and Hansen, R. H., *J. Adhes.*, **2** (1970), 93.
24. Yamakawa, S., *Macromolecules*, **12**(6) (1979), 1222.
25. Goldie, W., *Metallic coating of plastics*, Vol. 2, Electrochemical Publications Ltd, London 1969, p. 314.
26. Schonhorn, H., Frisch, H. L. and Gaines, G. L. Jr., *Polym. Eng.*, **17** (1977), 440.
27. Kinloch, A. J., *J. Mat. Sci.*, **15** (1980), 2141.
28. Brewis, D. M. and Briggs, D., *Polymer*, **22** (1981), 7.
29. Brockmann, W., *Adhäsion* (1978), (2), p. 38; *Adhäsion*, (3) p. 80; *Adhäsion* (4) p. 100.
30. Kinloch, A. J. in *Developments in adhesives*, W. C. Wake (Ed.), Applied Science, London 1977, p. 251.
31. Dwight, D. W., *J. Coll. Interface Sci.*, **59** (1977), 447.
32. Brecht, H., Mayer, F. and Binder, H., *Makromol. Chem.*, **33** (1973), 89.
33. Dwight, D. W. and Riggs, W. M., *J. Coll. Interface Sci.*, **47**(3) (1974), 650.
34. Jansta, J., Dousek, F. P. and Patzelova, V., *Carbon*, **13** (1975), 377.
35. Pelzbauer, Z., Baldrian, J., Jansta, J. and Dousek, F. P., *Carbon*, **17** (1979), 317.
36. Dousek, F. P., Jansta, F. P. and Baldrian, J., *Carbon*, **18** (1980), 13.
37. Jansta, J. and Dousek, F. P., *Electrochim Acta*, **18** (1973), 673.
38. Jansta, J. and Dousek, F. P., *Electrochim Acta*, **20** (1975), 1.
39. Brewis, D. M., Barker, D. J., Dahm, R. H. and Hoy, L. R. J., *Electrochim. Acta.*, **23** (1978), 1107.

40. Brewis, D. M., Barker, D. J., Dahm. R. H. and Hoy, L. R. J., *J. Mat. Sci.*, **14** (1979), 749.
41. Rifi, M. R. in *Techniques of electro-organic synthesis*, Vol. 5, Part II. N. L. Weinberg (Ed.), Wiley–Interscience, London, 1975.
42. Seiber, J. N., *J. Org. Chem.*, **36** (1971), 2000.
43. Paushkin, Ya. M., Vishnyakova, T. P., Lumin, A. F. and Nizova, S. A., *Organic polymeric semiconductors*, Wiley, New York, 1974. (Translated by R. Kondar, Translation Ed. D. Slutzkin).
44. Katon, J. E., *Organic semiconducting polymers*, Edward Arnold, London, 1968.
45. Barker, D. J. and Dahm. R. H., unpublished results.
46. Lagowski, J. J. and Moczygemba, G. A., in *The chemistry of non-aqueous solvents*, Vol. 2, J. J. Lagowski (Ed.), Academic Press, London, 1967, p. 319.
47. Simonet, J. and Lund, H., *J. Electroanal. Chem.*, **75** (1977), 719.
48. Besenhard, J. O. and Eichinger, G., *J. Electroanal. Chem.*, **72** (1976), 1.
49. Barker, D. J., Ph.D. Thesis, Leicester Polytechnic (1978).
50. Chiang, C. K., Druy, M. A., Gau, S. C., Heeger, A. J., Louis, E. J., MacDiarmid, A. G., Park, Y. W. and Shirakawa, H., *J. Amer. Chem. Soc.*, **100** (1978), 1013.
51. Mazur, S., Matusinovic, T. and Cammann, K., *J. Amer. Chem. Soc.*, **99** (1977), 3888.
52. Barker, D. J., Brewis, D. M., Dahm. R. H. and Hoy, L. R. J., *Polymer*, **19** (1978), 856.
53. Barker, D. J., Brewis, D. M., Dahm. R. H. and Hoy, L. R. J., *Extended Abstr. Internat. Conf. Adhesion and Adhesives: Science, Technology and Applications*, Gray College, Durham, UK, September 1980.
54. Weeks, N. E., Kohlmayer, G. M. and Otocka in *Characterization of metal and polymer surfaces*, Vol. 2, L.-H. Lee (Ed.), Academic Press, New York, 1977, p. 289.
55. Starkweather, H., *J. Polym. Sci., Polym. Phys. edn.*, **17** (1979), 73.
56. Schonhorn, H. and Ryan, F. W., *J. Adhes*, **1** (1969), 43.
57. Schonhorn, H. and Ryan, F. W., *J. Polym. Sci., Part A-2*, **7** (1969), 105.
58. Cadman, P. and Gossedge, G. M., *Wear*, **54** (1979), 211.
59. Schonhorn, H. and Roberts, F., *Coatings Plast. Prepr.*, **36**(2) (1977), 223.
60. Briggs, D. and Gribbin, J. D., unpublished results.
61. McBride, R. T. and Wolinski, L. F., US Patent 3,296,011 (1967).
62. Clark, D. T. and Feast, W. J., *J. Macromol. Sci., Revs. Macromol. Chem.*, **C12**(2) (1975), 191.
63. Clark, D. T. and Dilks, A., in *Characterization of metal and polymer surfaces*, L.-H. Lee (Ed.), Academic Press, New York 1977, p. 101.
64. Clark, D. T. and Dilks, A., *J. Polym. Sci., Polym. Chem. edn.*, **16** (1978), 911.
65. Smolinsky, G. and Vasile, M. J., *Eur. Polym. J.*, **15**(1) (1979), 87.
66. Yasuda, H. Y., Marsh, H. C. M. and Reilley, C. N. R., *J. Polym. Sci., Polym. Chem. edn.*, **15** (1977), 991.
67. Hirotsu, T. and Ohnishi, S., *J. Adhes.*, **11**(1) (1980), 57.
68. Kaelble, D. H., *Physical chemistry of adhesion*, Wiley–Interscience, New York, 1971.
69. Mashonov, A. and Avny, Y., *J. Appl. Polym. Sci.*, **25**(5) (1980), 771.

70. Brenman, M. and Lerchenthal, Ch.H., *Polym. Eng. Sci.*, **16** (1976), 747.
71. Brenman, M. and Lerchenthal, Ch.H., *Polym. Eng. Sci.*, **16** (1976), 760.
72. Yasaka, M., Japanese Patent 5,331,713 (1979); *Chem. Abs.*, **91**, 75473 m (1979).
73. Dahm, R. H. and Gribbin, J. D., unpublished results.
74. Meier, J. F. and Petrie, E. M., *J. Appl. Polym. Sci.*, **17** (1973), 1007.
75. Yamakawa, S., *Macromolecules*, **12**(6) (1979), 1222.
76. Galembeck, F., Galembeck, S. E., Vargas, H., Ribeiro, C. A., Miranda, L. C. M. and Ghizoni, C. C., *Chem. Abs.*, **92**, 94807 (1980).
77. Gribbin, J. D., unpublished results.

Chapter 11

SUMMARY AND CONCLUSIONS

D. M. Brewis

Leicester Polytechnic, Leicester, UK

Most important plastics and metals require a pretreatment to achieve satisfactory initial and long-term adhesion for printing, bonding and coating operations. Pretreatments have been developed, largely on an empirical basis, which usually give the required performance with the chosen substrate; however, on occasions unsatisfactory results are obtained. The treatments usually involve chemical change at the surface of the substrate, but with some materials in particular applications, physical pretreatments may be sufficient. Even where chemical changes have occurred, other factors may play an important role; for example, some believe that the success of phosphoric acid anodising of aluminium is due, at least in part, to the very rough surface which provides mechanical keying.

While it is true that the technology still leads the science of adhesion, the gap has been significantly narrowed in recent years. This is due partly to the large research effort, as exemplified by the work on aluminium, but in particular by some of the analytical techniques which have become available in recent years.

Most of the important methods which may be used to study plastic and metallic surfaces have been described in Chapters 2–6 and examples of their application are given in Chapters 7–10. Some of their features are given in Tables 1 and 2. Discussions and/or applications of other techniques have been given earlier in the book, e.g. SSIMS (Chapter 3), Rutherford backscattering (Chapter 3), Raman spectroscopy (Chapter 4), microprobe analysis (Chapter 5) and RHEED (Chapter 8). Several of the techniques have been used in surface studies for many years, whereas others have only become available in recent years.

TABLE 1

OUTLINE OF SOME IMPORTANT TECHNIQUES TO STUDY METALLIC SURFACES

Technique	Abbreviation	Information	Comments	Chapters
Optical microscopy		Surface topography and morphology	Inexpensive but modest resolving power and depth of field	5
Transmission electron microscopy	TEM	Surface topography and morphology	Very high resolution but requires replication—artefacts can be a serious problem	5, 8
Scanning electron microscopy	SEM	Surface topography and morphology—combined with X-ray spectroscopy gives 'bulk' elemental analysis	Resolving power $\gg$ optical microscopy. Preparation easier than TEM and artefacts much less likely	5, 8
Auger electron spectroscopy	AES	Chemical composition, depth profiling and lateral analysing	High spatial resolution which makes the technique especially suitable for composition–depth profiling	2, 3, 7, 8
X-ray photoelectron spectroscopy[a]	XPS	Chemical composition—depth profiling	Especially useful for studying the adhesion of polymers to metals	2, 7, 8
Secondary ion mass spectroscopy	SIMS	Elemental analysis in 'monolayer range'—chemical composition and depth profiling	Extremely high sensitivity for many elements	3, 8
Contact angle measurement	θ	Contamination by organic compounds	Inexpensive; rapid	6

[a] Also called Electron Spectroscopy for Chemical Analysis, ESCA.

TABLE 2
OUTLINE OF SOME IMPORTANT TECHNIQUES TO STUDY PLASTIC SURFACES

Technique	Abbreviation	Information	Comments	Chapters
Optical microscopy		Surface topography and morphology	Inexpensive but modest resolving power	5
Transmission electron microscopy	TEM	Surface topography and morphology	Very high resolution but requires replication—artefacts can be a serious problem	5
Scanning electron microscopy	SEM	Surface topography and morphology	Resolving power $>>$ optical microscopy Preparation easier than TEM and artefacts much less likely	5
X-ray photelectron spectroscopy	XPS	Elemental analysis and nature of chemical groups	Little, if any, damage to polymers, unlike AES.	4, 9
Multiple internal reflection—IR	MIR-IR	Chemical groupings and morphology	Less surface sensitive than XPS but FT-IR can give a significant improvement.	4, 9
Contact angle measurement	θ	Surface energies—contributions of dispersion and polar forces and H-bonds to surface energy	Inexpensive equipment but great care required with experimental work	6, 9

To obtain a good understanding of the effects of a pretreatment, it is necessary to use at least two techniques and preferably more. In particular, it is necessary to follow changes in topography, e.g. with SEM, and chemical changes, e.g. with ESCA. The techniques available are often complememtary; in combinations such as AES/SIMS, ISS/SIMS and XPS/SIMS, each technique can utilise the UHV facilities.

The costs of the techniques range from a few pounds for contact angle measurements to more than £ 100 000 for ESCA/SIMS. The *approximate* order of costs is:

contact angles < optical microscopy < SEM ~ TEM ~
MIR–IR < AES ~ ESCA ~ SIMS ~ FT–IR

Some of the newer techniques have already made a large contribution to the understanding of adhesion problems involving plastic and metallic substrates (see in particular Chapters 2, 3, 4, 8 and 9). For example, the evidence in favour of the weak boundary explanation for the poor adhesion of polyolefins has been much weakened. Various workers, having failed to detect oxidation after pretreatments, concluded that weak boundary layers had been eliminated. Thus Schonhorn and Ryan[1] concluded that weak boundary layers had been eliminated by the formation of a transcrystalline region after melting PE against aluminium; likewise Blais *et al.*[2] concluded that the chromic acid treatment of PP was effective by removing a weak boundary layer. However, later studies of these treatments using XPS showed that substantial oxidation had occurred with both these treatments as described in references 3 and 4 respectively. XPS has also provided evidence indicating the importance of specific groups (Chapter 9), but more work is required to understand the interaction between groups introduced by pretreatments and the adhesive, coating or ink.

In the case of PTFE the situation is clearly more complex. There is evidence for the transfer, at low loads, of fluorinated material from the PTFE to the adhesive. However, the most widely used pretreatment involves the formation of a distinct carbonaceous layer whose surface possesses various oxygen-containing groups; it is not clear whether these groups are vital for good adhesion. Other complications associated with PTFE are discussed in Chapter 10.

Although the more recent methods of chemical analysis have made a big impact on adhesion studies, the more traditional methods still have an important role to play. This was demonstrated in a study of the corona treatment of poly(ethylene terephthalate) by Briggs *et al.*[5] in which XPS and contact angle measurements provided complementary information.

With metals, the key problems involve the relationships between surface treatments, trace impurities and durability, both for coatings and adhesive joints (see Chapters 3, 7 and 8). Various combinations of analytical techniques have already provided much information on these relationships but clearly more work is required for the detailed understanding which will greatly enhance durability. There is still controversy regarding the mechanisms by which water can adversely affect adhesion (displacement of polymer, corrosion, formation of weak oxides, etc.) and again the modern analytical methods are providing useful information in this area (see Chapter 8 and reference 6).

More stringent environmental regulations combined with the increasing use of structural adhesives, e.g. in road vehicles, will necessitate alternatives to chromic acid. The effects of any new treatments must be carefully studied. Demanding applications that exist in the aerospace and other industries will also require further detailed investigations especially as new materials or modifications are introduced. The range of techniques now available should lead to a successful outcome of such investigations. However, as noted earlier (e.g. Chapter 3 *re.* SIMS and Chapter 5 *re.* TEM), great care must be exercised to avoid misinterpretation of results.

REFERENCES

1. Schonhorn H. and Ryan F. W., *J. Appl. Polym. Sci., Part A–2*, **6** (1968), 231.
2. Blais P., Carlsson D. J., Csullog G. W. and Wiles D. M., *J. Coll. Interface Sci.*, **47**(3) (1974), 636.
3. Briggs D., Brewis D. M. and Konieczko M. B., *J. Mat. Sci.*, **12** (1977), 429.
4. Briggs D., Brewis D. M. and Konieczko M. B., *J. Mat. Sci.*, **11** (1976), 1270.
5. Briggs D., Rance D. G., Kendall C. R. and Blythe A. R., *Polymer*, **21** (1980), 895.
6. Kinloch A. J., 19th Annual Conference on Adhesion and Adhesives, City University, London, 31 March—1 April 1981, to be published in *Adhesion—6* by Applied Science Publishers, London.

APPENDIX: MAIN ABBREVIATIONS USED

MATERIALS

ABS	acrylonitrile–butadiene–styrene
FEP	fluorinated ethylene–propylene
HDPE	high density polyethylene
LDPE	low density polyethylene
PB	polybutadiene
PE	polyethylene
PET	poly(ethylene terephthalate)
PP	polypropylene
PTFE	polytetrafluoroethylene
PVdF	poly(vinylidene fluoride)
PVF	poly(vinyl fluoride)

TECHNIQUES

AES	Auger electron spectroscopy
ATR-IR	attenuated total reflectance infrared
EDAX	energy dispersive microanalysis
ESCA	electron spectroscopy for chemical analysis
FT-IR	Fourier-transform infrared
ISS	ion scattering spectroscopy
MIR-IR	multiple internal reflection infrared
RHEED	reflection high energy electron diffraction
SCANIIR	surface composition by analysis of neutral ion impact radiation

SEM	scanning electron microscopy
SIMS	secondary ion mass spectroscopy
SSIMS	static SIMS
STEM	scanning transmission electron microscopy
TEM	transmission electron microscopy
XPS	X-ray photoelectron spectroscopy

MISCELLANEOUS

BE	binding energy
CRT	cathode ray tube
DSC	differential scanning calorimetry
FPL	Forest Products Laboratory
fsd	full scale deflection
IMFP	inelastic mean free path
KE	kinetic energy
NA	numerical aperture
pbw	parts by weight
RH	relative humidity
S/N	signal/noise
T_g	glass transition temperature
UHV	ultra-high vacuum
VB	valence band
WD	wavelength dispersive
γ_c	critical surface tension
θ	contact angle or 'take-off' angle

INDEX